1

Second Arctic Expedition

Part One: Narrative; Western Detachment

John Franklin, R.N., F.R.S.

Published by:
CD-Academia Book Company,
BEDFORD, NS.
860 Shore Drive,
CANADA, B4A 2E4

NARRATIVE OF A SECOND EXPEDITION

TO THE SHORES OF

THE POLAR SEA,

IN THE YEARS

1825, 1826, AND 1827,

BY

JOHN FRANKLIN, Captain R.N., F.R.S., &c.

AND COMMANDER OF THE EXPEDITION.

INCLUDING AN ACCOUNT OF THE PROGRESS OF A DETACHMENT
TO THE EASTWARD,

BY

JOHN RICHARDSON, M.D., F.R.S., F.L.S., &c.

SURGEON AND NATURALIST TO THE EXPEDITION.

ILLUSTRATED BY NUMEROUS PLATES AND MAPS.

---

*PUBLISHED BY AUTHORITY OF THE RIGHT HONOURABLE THE SECRETARY OF STATE
FOR COLONIAL AFFAIRS.*

LONDON:

JOHN MURRAY, ALBEMARLE-STREET.

MDCCCXXVIII.

LONDON :
PRINTED BY WILLIAM CLOWES,
Stamford-Street.

TO

THE RIGHT HONOURABLE

THE EARL BATHURST, K.G.,

LORD PRESIDENT OF HIS MAJESTY'S COUNCIL,

&c. &c. &c.

THE FOLLOWING

NARRATIVE OF A SECOND JOURNEY OF DISCOVERY ALONG
THE NORTHERN COAST OF AMERICA,

UNDERTAKEN BY ORDER AND UNDER THE AUSPICES OF
HIS LORDSHIP,

IS, BY PERMISSION, INSCRIBED,

WITH GREAT RESPECT AND GRATITUDE,

BY

THE AUTHOR.

CONTENTS.

CONTENTS.

Dr. Richardson's Narrative of the Proceedings of the Eastern Detachment of the Expedition.

Captain Franklin's Narrative resumed.

APPENDIX.

An Account of the objects of Natural History, collected on our journey, being too voluminous to be inserted in the Appendix, has been reserved for a separate Work, which will be published as soon as possible, by Dr. Richardson and Professor Hooker, under the sanction, and by the assistance, of His Majesty's Government.

LIST OF PLATES.

Maps at the end of the Book in the following order:

* See Appendix, Part V., page cxxx, for a notice respecting the longitude of Fort William. The Maps 1, 2, 3, contain the routes which are followed by the Fur Traders in going from Canada and Hudson's Bay to the Northern Departments. The two latter were constructed on the former Expedition, and are now used, as we pursued the course therein delineated from the Saskatchawan River to Slave Lake.

INTRODUCTORY CHAPTER.

His Majesty's Government having, towards the close of the year 1823, determined upon another attempt to effect a northern passage by sea between the Atlantic and Pacific Oceans, and Captain Parry, the highly distinguished Commander of the two preceding Expeditions, having been again entrusted with its execution, success, as far as ability, enterprise, and experience could ensure it, appeared likely to be the result. Yet, as the object was one for which Great Britain had thought proper to contend for upwards of three centuries, it seemed to me that it might be desirable to pursue it by more ways than one; I therefore ventured to lay before His Majesty's Government a plan for an Expedition overland to the mouth of the Mackenzie River, and thence, by sea, to the north-western extremity of America, with the combined object, also, of surveying the coast between the Mackenzie and Coppermine Rivers.

I was well aware of the sympathy excited in the British public by the sufferings of those engaged in the former overland Expedition to the mouth of the Coppermine River, and

b

of the humane repugnance of His Majesty's Government to expose others to a like fate; but I was enabled to show satisfactorily that, in the proposed course, similar dangers were not to be apprehended, while the objects to be attained were important at once to the naval character, scientific reputation, and commercial interests of Great Britain; and I received directions from the Right Honourable Earl Bathurst to make the necessary preparations for the equipment of the Expedition, to the command of which I had the honour to be nominated.

My much valued friend, Dr. Richardson, offered his services as Naturalist and Surgeon, and also volunteered to undertake the survey of the coast between the Mackenzie and Coppermine Rivers, while I should be occupied in endeavouring to reach Icy Cape.

Lieutenant Bushnan, who had served under Captains Ross and Parry on their voyages of discovery, was also appointed to accompany me; but, long before the party was to leave England, I had to lament the premature death of that excellent young officer, who was eminently qualified for the situation, by his skill in astronomical observations, surveying, and drawing. Many naval officers, distinguished for their talent and ability, were desirous of filling the vacancy; but my friend and former companion, Lieutenant Back, having returned from the West Indies, the appointment was offered to him, and accepted with his wonted zeal.

Mr. E. N. Kendall, Admiralty Mate, and recently Assistant

Surveyor with Captain Lyon, was appointed to accompany Dr. Richardson in his voyage to the eastward, and to do the duty of Assistant-Surveyor to the Expedition at large, whilst it continued united. Lastly, Mr. Thomas Drummond, of Forfar, was appointed Assistant Naturalist, on the recommendation of Professor Hooker and other eminent scientific men.

A residence in the northern parts of America, where the party must necessarily depend for subsistence on the daily supply of fish, or on the still more precarious success of Indian hunters, involves many duties which require the superintendence of a person of long experience in the management of the fisheries, and in the arrangement of the Canadian voyagers and Indians : we had many opportunities, during the former voyage, of being acquainted with the qualifications of Mr. Peter Warren Dease, Chief Trader of the Hudson's Bay Company, for these services, and I therefore procured the sanction of His Majesty's Government for his being employed on the Expedition.

As soon as I had authority from Earl Bathurst, I entered into a correspondence with the Governor and Directors of the Hudson's Bay Company; and these Gentlemen, taking the most lively interest in the objects of the Expedition, promised their utmost support to it, and forthwith sent injunctions to their officers in the Fur Countries to provide the necessary depôts of provision at the places which I pointed out, and to give every other aid in their power. I also wrote to the different Chief Factors and Chief Traders of the Company, who resided on the route of the Expedition, explaining its objects, and requesting their co-operation.

Pemmican, the principal article of provision used in travelling, being made during the winter and spring, the orders for providing the extra quantity required for the Expedition, though sent out from England by the earliest conveyance, so as to reach the provision posts in the summer of 1824, could not be put into effect sooner than the spring of 1825; hence, it was not proper that the main body of the Expedition should reach the Fur Countries before the latter period. Some stores were forwarded from England, by way of New York, in March 1824, under charge of Mr. Robert M'Vicar, Chief Trader, for the purpose of relieving the Expedition as much as possible from the incumbrance of heavy baggage, and thus enabling it, by marching quickly, to reach its intended winter-quarters at Great Bear Lake, as well as to provide for its more comfortable reception at that place. These stores, with the addition of other articles obtained in Canada, sufficed to load three north canoes, manned by eighteen voyagers; and they were delivered by Mr. M'Vicar, before the winter set in, to Mr. Dease, at the Athabasca Lake. Mr. Dease was instructed to support his party by fishing at Great Slave Lake, during the winter of 1824-25; and, early in the spring of 1825, to proceed to Great Bear Lake, and commence the necessary buildings for the reception of the Expedition. I may here cursorily remark that, in selecting Great Bear Lake as our winter residence, I was influenced by the information I had obtained of its being the place nearest to the mouth of the Mackenzie, known to the traders, where a sufficient supply of fish could be procured for the support of so large a party.

Three light boats, which I shall soon more particularly describe, were also sent out to York Factory, in June 1824, in the annual Hudson's Bay ship, together with a further supply of stores, two carpenters, and a party of men, with a view of their reaching Cumberland House, on the Saskatchawan River, the same season; and, starting from thence as soon as the navigation opened in the following spring, that they might be as far as possible advanced on their way to Bear Lake before they were overtaken by the Officers of the Expedition. The latter, proceeding by way of New York and Canada, would have the advantage of an earlier spring in travelling through the more southern districts; and, further to expedite their progress, I directed two *large* canoes (canôts de maître), with the necessary equipments and stores, to be deposited at Penetanguishene, the naval depôt of Lake Huron, in the autumn of 1824, to await our arrival in the following spring; having been informed that, in ordinary seasons, we should, by commencing our voyage at that place, arrive in the north-west country ten days earlier than by the usual way of proceeding up the Utawas River from Montreal.

The return of the Hudson's Bay ship towards the close of the year 1824, brought me satisfactory intelligence of the progress of the above-mentioned parties, together with the most pleasing assurances from the Gentlemen of the Company to whom I had written, of their zeal in our cause; and here I must express the deep sense I have of the kindness of the late Honourable William M'Gillivray, of Montreal, whose experience enabled him to give me many valuable suggestions relating to the clothing and subsistence of the party, and to the supplies proper for the Indians.

In connexion with the above sketch of the preparatory steps taken in the course of the year 1824, it may be proper to give, in this place, a short account of the general equipments of the Expedition.

And first, with regard to the vessels intended for the navigation of the Arctic Sea: birch-bark canoes, uniting lightness and facility of repair with speed, are certainly well adapted for navigating the rivers of America, but they are much too slight to bear the concussion of waves in a rough sea, and they are still less fitted, from the tenderness of the bark, for coming in contact with ice. I therefore requested of the Lords Commissioners of the Admiralty that *three* boats might be constructed under my superintendence; and they were immediately ordered and promptly finished under the directions of the Commissioners of the Navy. To fit them for the ascent and descent of the many rapids between York Factory and Mackenzie River; and to render their transport over the numerous portages more easy, it was necessary to have them as small, and of as light a construction as possible; and, in fact, as much like a north canoe as was consistent with the stability and capacity required for their voyage at sea. They were built of mahogany, with timbers of ash, both ends exactly alike, and fitted to be steered either with a sweep-oar or a rudder. The largest, twenty-six feet long, and five feet four inches broad, was adapted for six rowers, a steersman, and an officer; it could be borne on the shoulders of six men, and was found, on trial, to be capable of carrying three tons weight in addition to the crew. The two others were each twenty-four feet long, four feet ten inches broad, and were capable of receiving a crew of five men, a steersman, and an officer, with an additional weight of two and a half tons. The greatest care was paid to their construction by Mr. Cow, boat-builder of Woolwich Yard; and, as I could not often be present, my friend Captain Buchan, R.N., kindly undertook to report their progress; and I am further indebted to him for many valuable suggestions which were acted upon.

When the boats were finished, they were tried at Woolwich, in the presence of many naval and military officers, as to their qualities of sailing, rowing, and paddling, and found to answer fully the expectations that had been formed of them. At the same time we tried another little vessel belonging to the Expedition, named the Walnut-Shell, the invention and construction of which I owe to my friend Lieutenant-Colonel Pasley, of the Royal Engineers. Its length was nine feet, its breadth four feet four inches, and it was framed of well-seasoned ash, fastened with thongs, covered with Mr. Mackintosh's prepared canvas, and shaped like one valve of a walnut-shell, whence its appellation. It weighed only eighty-five pounds, could, when taken to pieces, be made up in five or six parcels, and was capable of being put together in less than twenty minutes. So secure was this little vessel, that several ladies, who had honoured the trial of the boats with their presence, fearlessly embarked in it, and were paddled across the Thames in a fresh breeze. It was intended to provide against a similar detention in crossing rivers to that which proved so fatal to our party on the former journey; and it was also thought, that this little bark would be found useful in procuring water-fowl on the small lakes, to which the boats could not be conveyed.

In the choice of astronomical instruments I was necessarily guided by their portability. Our stock consisted of two small sextants, two artificial horizons, two altitude instruments, a repeating circle for lunar observations, and a small transit telescope for ascertaining the rates of the chronometers. We had a dipping-needle mounted on Meyer's plan, a plain needle very delicately fitted for observing the diurnal variation; two of Kater's azimuth compasses, and a pocket compass for each officer. The atmospherical instruments were two electrometers, two of Daniel's hygrometers, Leslie's photometer and hygrometer, besides a good supply of mercurial and spirit thermometers of different sizes. The magnetic instruments were examined in concert with my friend Captain Sabine, previous to my departure from London; and the observations that were obtained for dip and intensity, served as points of comparison for our future results.

The stores consisted of bedding and clothing, including two suits of water-proof dresses for each person, prepared by Mr.

Mackintosh, of Glasgow; our guns had the same bore with the fowling-pieces, supplied by the Hudson's Bay Company to the Indian hunters, that is, twenty-eight balls to the pound; their locks were tempered to withstand the cold of the winter; and a broad Indian dagger, which could also be used as a knife, was fitted to them, like a bayonet. Ammunition of the best quality was provided by the Ordnance, the powder being secured in small field or boat magazines. A quantity of wheaten-flour, arrow-root, macaroni, portable-soup, chocolate, essence of coffee, sugar, and tea, calculated to last two years, was also supplied, made up into packages of eighty-five pounds, and covered with three layers of prepared waterproof canvas, of which material coverings for the cargo of each boat were also made.

There was likewise an ample stock of tobacco, a small quantity of wine and spirits, marquees and tents for the men and officers, some books, writing and drawing paper, a considerable quantity of cartridge-paper, to be used in preserving specimens of plants; nets, twine, fishing-lines and hooks, together with many articles to be used at winter-quarters, for the service of the post, and for the supply of our Indian hunters, such as cloth, blankets, shirts, coloured belts, chiefs' dresses, combs, looking-glasses, beads, tapes, gartering, knives, guns and daggers, hatchets, awls, gun-worms, flints, fire-steels, files, whip and hand saws, ice-chisels and trenching-irons, the latter to break open the beaver lodges.

As the mode of travelling through the Hudson's Bay territories, with all its difficulties and hazards, is now well known to the

22

c

public, I think it better to give in this Introductory Chapter a slight outline of our route through the United States, Upper Canada, and Southern part of the Fur Countries, and to commence the detailed Narrative of the proceedings of the Expedition with its arrival in Methye River, where the officers joined the boats that had been sent out from England in the preceding year.

On the 16th of February, 1825, I embarked with Lieutenant Back, Dr. Richardson, Mr. Kendall, Mr. Drummond, and four marines, at Liverpool, on board the American packet-ship, Columbia, Captain Lee; and, on quitting the pier, we were honoured by a salute of three animating cheers, from a crowd of the principal inhabitants, who had assembled to witness our departure. The passage across the Atlantic was favourable and pleasant, and our reception at New York kind in the extreme. We landed at that city on the 15th of March, and our baggage and stores were instantly passed through the Custom-House without inspection. Cards of admission to the Public Scientific Institutions were forwarded to us the same evening, and, during our stay every other mark of attention was shown by the civil and naval authorities, as well as by private individuals, indicating the lively interest which they took in our enterprise.

James Buchanan, Esq., the British Consul, in addition to many other attentions, kindly undertook to accommodate a journey he had to make to Upper Canada, so as to accompany us through the State of New York. After a stay of eight days in the city, for the purpose of obtaining the rates of the chronometers, and for making some other observations with Meyer's dipping needle, we embarked, under the Consul's guidance, in the steam-boat Olive Branch, and ascended the Hudson River, to Albany, where we experienced similar civilities to those we had received at New York. Every body seemed to desire our success, and a fervent prayer for our preservation and welfare was offered up by the Reverend Dr. Christie, the minister of the church that we attended. The Honourable De Witt Clinton, the Governor of the State, assured me, that had we not been accompanied by a gentleman so conversant in the different routes and modes of travelling as Mr. Buchanan, he would have sent his son with us, or would himself have conducted us to the confines of the State.

From Albany, we travelled through Utica, Rochester, and Geneva, to Leweston, in coaches, with more or less rapidity, according to the condition of the roads; and, crossing the river Niagara, entered Canada, and visited the Falls so justly celebrated as the first in the world for grandeur. We next crossed Lake Ontario in a sailing boat, and came to York, the capital of Upper Canada, where we were kindly received by the Lieutenant-Governor Sir Peregrine Maitland, and by Colonel Cockburn and the Commissioners then employed on an inquiry respecting the value of the Crown Lands. From York we passed on to Lake Simcoe, in carts and other conveyances, halting for a night at the hospitable house of Mr. Robinson of Newmarket. We crossed Lake Simcoe in canoes and boats, and landed near the upper part of Kempenfeldt Bay, but not without being obliged to break our way through the ice for a short distance. A journey of nine miles, performed on foot, brought us to the River Nattawassaga, which we descended in a boat; and passing through a part of Lake Huron, arrived at Penetanguishene. At this place, we were hospitably entertained by Lieutenant, now Captain Douglass, during eight days that we waited for the arrival of our Canadian voyagers from Montreal.

We left Penetanguishene on St. George's day (23d April) in the two large canoes, which had been deposited at that place in the preceding autumn, our party, by the accession of the voyagers, now amounting to thirty-three; and after a few days, detention by ice, and bad weather, we reached Sault de St. Marie on the 1st of May, being ten days or a fortnight earlier than the oldest resident remembered a canoe from Canada to have arrived. From the Sault de St. Marie, we coasted the northern shore of Lake Superior to Fort William, formerly the great depôt of the N.W. Company, where we arrived on the 10th of May. We now exchanged our two *canôts de maître* for four small north canoes, in one of which, more lightly laden, Dr. Richardson and I embarked, with the view of proceeding as rapidly as possible to arrange supplies of provision at the different posts, while Lieutenant Back was left to bring up the three remaining and more deeply laden canoes.

We proceeded by the route delineated in the maps through Rainy Lake, the Lake of the Woods, Lake Winipeg, and the Saskatchawan River to Cumberland House, where we arrived on the 15th of June, and learned that our boats had left that place on the 2d of the same month. We found, also, with deep regret, that Thomas Mathews, the principal carpenter, who had accompanied the boats from England, had had the misfortune to break his leg the evening before their departure. But, fortunately, an officer of the Hudson's Bay Company then present, had sufficient skill to set it, and Dr. Richardson now pronounced that in two months he would be able to come on in one of the Company's canoes, and join us at the Bear Lake, which he was very desirous of doing. I therefore made arrangements to this effect, and also concerning supplies for Mr. Drummond the Assistant Naturalist, who was to be employed, during our stay in the north, in making collections in the vicinity of the Rocky Mountains.

Having remained one night at Cumberland House, we resumed our voyage, and passing through Pine Island Lake, Beaver Lake, crossing the Frog Portage, and ascending the English River, with its dilatations, named Bear Island, Sandfly, Serpent, Primeau, and Isle à la Crosse Lakes, we came to the post situated on, and named from the latter sheet of water, at four P.M. on the 25th June. In the course of this voyage, we met the Gentlemen of the Hudson's Bay Company proceeding from the interior with various brigades of canoes, carrying the returns of trade for the year to York Factory, and I had not only the satisfaction of hearing frequent news of the progress of our boats, but that the deposits of provisions I had requested, and the other arrangements I had made, were all punctually carried into effect. Mr. Spencer, the gentleman in charge at

Isle à la Crosse, informed us, that the boats had gone off a few hours previous to our arrival, with the addition of a batteau laden with pemmican, under the charge of Mr. Fraser, a clerk of the Hudson's Bay Company.

I waited at this establishment one night to obtain astronomical observations, and to bespeak an additional quantity of provisions, &c., which being satisfactorily done, we resumed our voyage on the 27th, and, passing through Deep River, Clear and Buffalo Lakes, overtook the boats in Methye River, at sunrise on the 29th of June.

Having brought this preliminary sketch up to the date at which the ensuing Narrative of the proceedings of the Expedition commences, I turn to the pleasing duty of rendering my best thanks to the many gentlemen who have assisted me in forwarding its progress. To the Right Honourable Earl Bathurst, I am greatly indebted for the readiness with which he attended to every suggestion I had to make regarding the equipment of the Expedition, and to the Right Honourable Wilmot Horton, the Under Colonial Secretary, for his kindness and promptitude in facilitating all my views. Nor can I feel less grateful to Lord Viscount Melville, and to the Lords Commissioners of the Admiralty for their patronage and support, as well as to Sir Byam Martin, the Comptroller, and to the Commissioners of the Navy and Victualling Offices, for the arrangements depending on their boards. Mr. Pelly, the Governor of the Hudson's Bay Company, and Mr. Garry, the Deputy-Governor, as well as every Member of its Committee, claim my most sincere thanks for their unremitting endeavours to promote the welfare of the Expedition through its whole progress; and I feel truly obliged to Mr. Simpson, the Governor in the Fur Countries; to Mr. M'Tavish, Mr. Haldane, Mr. M'Donald, Mr. Leith, Mr. Stuart, and Messrs. James and George Keith, Chief Factors, who, acting in the spirit of their instructions, were very assiduous in collecting provisions and stores for the use of my party, and in forwarding all our supplies. There were other gentlemen resident in the more northern parts of the country, to whom I am no less obliged for advice and assistance; but the brevity requisite in this place necessarily compels me to refer to the Narrative, where their names, and the services they rendered, are mentioned.

I cannot, however, close this Introductory Chapter, without expressing the deepest obligation to those kind friends and excellent officers with whom I had the happiness of being associated, who constantly aided me by their most cordial co-operation, and whose best efforts were devotedly applied to every pursuit which could be interesting to science. Nor can I omit to mention the gratitude I owe to each of the seamen, marines, British and Canadian voyagers who composed our party at the winter-quarters, for their steady obedience and truly good conduct, whether in the days of relaxation during the winter, or in the more arduous exertions of our summer occupations.

The Views with which the ensuing Narrative is illustrated, have been chosen from a large collection of drawings, made by Captain Back and Lieutenant Kendall. Of the interesting nature of the scenes which have been selected, the reader will be the best judge ; but for their fidelity, I pledge myself in the most unreserved manner.

_____

OFFICIAL INSTRUCTIONS.

————————

Downing-street, 31st Jan., 1825.

Sir,

His Majesty's Government having decided that an Expedition should be set forth, for the purpose of exploring the Northern Coast of America, between the Mouth of Mackenzie's River, and the Strait of Behring; and confiding in your zeal and experience for the due execution of this service, I have recommended you as a proper person to be charged with the same. You are, therefore, to proceed with your party (a list of whom is annexed) by the Packet from Liverpool to New York, and from thence make the best of your way to Lake Huron, where the stores necessary for your journey have already been sent. Embarking in Canoes, you are from thence to follow the water communication to the western side of the Great Bear Lake, where you are to establish your winter-quarters; and having so done, your first care should be to endeavour to open a friendly communication with the Esquimaux.

Early in the Spring of 1826, you are to proceed down the Mackenzie River with all the necessary stores and provisions, in order to be prepared to take advantage of the first opening of

d

the ice on the Polar Sea, so as to enable you to prosecute your voyage along the coast to Icy Cape, round which you are to proceed to Kotzebue's Inlet, where you may expect to find His Majesty's Ship, Blossom, which the Lords Commissioners of the Admiralty will order to proceed to that rendezvous, in the Summer of 1826. But if, on your arrival at Icy Cape, or the northern point of Behring's Strait, you should be of opinion that you could, with safety, return the same season to the established winter-quarters, you are at liberty to do so, instead of proceeding to join the Blossom. You will, therefore, without loss of time, settle with Captain Beechey, her commander, such a plan as may appear to you, both, best adapted for ensuring your meeting together, and establish a code of signals, or devise such other means as may tend to give you information, if possible, previous to your reaching the longitude of Icy Cape.

On your arrival at the mouth of Mackenzie River, you are to despatch Dr. Richardson with Mr. Kendall and five or six men, in one of the boats, to examine the intermediate coast between the Mackenzie and Coppermine Rivers; but if you should find that the stores and provisions you have been able to accumulate are not sufficient for your own and Dr. Richardson's party, you are, in that case, to direct Dr. Richardson to employ himself and party on shore, in examining the country contiguous to the Mackenzie River, the Rocky Mountains, the shores of the Great Bear Lake, the Copper Mountains, and as far round as he can with safety, collecting specimens of the animals, plants and minerals, and also laying in a stock of provisions sufficient for both parties, if, by any unforeseen accident, you should find yourself compelled to return without reaching the Blossom.

If, in proceeding westerly towards Icy Cape, you should make but slow progress, and find yourself impeded by ice or land jutting out to the northward farther than is calculated upon, or from accidents to the boats, or any other unforeseen circumstance, so that it remains doubtful whether you will be able to reach the neighbourhood of Kotzebue's Inlet the same season, you are not to consider yourself authorized to risk yourself and party to the chance of being obliged to winter on the coast, but commence your return about the 15th or 20th of August to the established winter-quarters on Bear Lake, unless you should be satisfied that yourself and party could pass the winter with safety among the Esquimaux, and that there was afforded a certainty of your reaching Behring's Strait the following Season' when the Blossom will again proceed to the appointed rendezvous.

In the event of your reaching Kotzebue's Inlet the first season, Captain Beechey will be instructed to convey you and your party in the Blossom to the Sandwich Islands or Canton, as may seem most advisable to you, from whence you will be able to take a passage to England in one of the Company's Ships or Private Traders; and you will leave such instructions with Dr. Richardson for his guidance, in the event of your being able to accomplish this point, as you may deem fit and proper for his return to England.

In the event of your death, or any accident which may prevent your proceeding, the command of the Expedition must necessarily devolve on Lieutenant Back, who is to follow these Instructions; but he is not to alter any arrangement with regard to Dr. Richardson's proceedings which you may have settled for him to pursue, the principal object of Dr. Richardson's accompanying you, being that of completing, as far as can be done, our knowledge of the Natural History of North America. Lieutenant Back will, therefore, in the event abovementioned, act in concert with Dr. Richardson, but not direct him and his party from any plan of operations which he and you may previously have settled.

You will take care to inform me from time to time, as opportunities may occur, of your proceedings, and the progress made in the Expedition, with the direction of which you are hereby entrusted.

I have the honour to be,

Sir,

Your most obedient Servant,

BATHURST.

To Captain Franklin, R.N.,
     &c. &c. &c.

SECOND JOURNEY TO THE SHORES

OF

THE POLAR SEA.

CHAPTER I.

Join the Boats in the Methye River—Cross the Long Portage—Arrival at Fort Chipewyan —Departure from thence with the whole party for Mackenzie River—Arrangements at Fort Norman—Descent to the Sea—Return to the Winter Quarters at Great Bear Lake.

THE boats of the Expedition had advanced from Hudson's Bay into the interior, twelve hundred miles, before they were joined by the officers; whilst the latter, from taking a more circuitous route by New York and Canada, as shewn in the introductory chapter, travelled two thousand and eight hundred miles, to reach the same point.

This junction took place early in the morning of the 29th of June, 1825, in the Methye River, latitude 56° 10′ N., longitude 108° 55′ W., which is almost at the head of the waters that flow from the north into Hudson's Bay.

In no part of the journey was the presence of the officers more requisite to animate and encourage the crews, because the river itself, beside being obstructed by three impassable rapids, is usually so shallow, through its whole course of forty miles, as scarcely to admit of a flat-bottomed bateau floating with half its

B

1825.
June.

cargo, much less our boats, which drew, when loaded, from eighteen to twenty inches.   This river and its impediments being surmounted, the Methye Portage, ten miles and three quarters long, was at no great distance, which is always held up to the inexperienced voyager as the most laborious part of the journey. But whatever apprehensions the men might have entertained on this subject, seemed to vanish on our landing amongst them; and Dr. Richardson and myself were received by all with cheerful, delighted countenances, and by none more warmly than by our excellent friend and former interpreter Augustus the Esquimaux, and Ooligbuck, whom he had brought from Churchill, as his companion.   A breakfast was quickly prepared by Mr. Fraser, a clerk of the Hudson's-Bay Company, under whose charge the boats had been, since their departure from Cumberland House; and I then inspected the boats and stores, which I was rejoiced to find were in good order.   We had brought letters from the relatives of several of the party, and another hour was allowed to read them.

At ten A.M. we began to ascend the stream, but very soon found that it was necessary for the whole party to walk in the water, and drag the boats through the mud.   Nor could we long advance even by this mode, but were compelled either to carry some of the cargo along the shore, where walking was at all practicable, or else to take half the lading in a boat to a part where the river was deeper, and then return for the remainder.   From thus travelling the distance twice over, it was the fifth day before we reached the lake from whence its waters flow.

Thursday,
June 30.

On the evening of the 30th of June, we witnessed one of those violent but momentary gusts of wind which occur not unfrequently in the spring and autumn, and which prove so destructive to the forests in this country.   It was preceded by calm and very sultry weather, with loud thunder and vivid lightning.   In an instant, the tents were overthrown, and even very large trees were bent by its force into a horizontal position; indeed, for a few seconds, the scene around us appeared one of almost entire devastation. When the violence of the squall was past, we had great reason to be pleased at its occurrence, for the strong steady breeze and heavy rain that succeeded, carried away the myriads of musquitoes by which we had been tormented the whole day.

1825.
June.

Having crossed the Methye Lake, we arrived at the portage of that name. Here it was necessary to make an equal division of the cargoes, and to devise means for the conveyance of the boats. The packages amounted to one hundred and sixteen, weighing from seventy to ninety pounds each, exclusive of the three boats and the men's personal luggage; and there were nineteen men of the boats' crews, two Canadians, and two boys, to carry these burdens. At first the packages were equally distributed among this party; but several of the men, who had been reduced by their previous exertions, became lame: among these were the Esquimaux, and we were, therefore, compelled to make other arrangements, and ultimately to employ the crew of my canoe, though the great fatigue they had suffered in our rapid journey from Penetanguishene, made me desirous of sparing them for the present.

Monday,
July 4.

The boats were the heaviest and most difficult articles to transport. One of the small boats was carried on the shoulders of eight men, of whom Mr. Fraser undertook to be one, as an example to the rest. Another of the same size was dragged by other eight men; and the largest was conveyed on a truck made for the purpose on the spot, to which service the lame were attached.

Each day's journey, and also the intermediate stages, were determined by the places where water could be procured, and our mode of travelling was as follows:—Rising at three A.M., the

B 2

men carried a part of their burden to the first stage, and continued to go backwards and forwards till the whole was deposited. They then slept for a few hours, and in the cool of the evening the boats were brought up. By these means every thing was ready at the western end of the portage early on Monday, the

Monday,
11th.
11th of July. The slight injuries which the boats had received, principally from exposure to the sun, were soon repaired; they were put into the water to tighten, and the whole party were allowed to rest.

With reference to the Methye Portage I may remark, that, except the steep hill at its western extremity, the road is good and tolerably level, and it appeared to us that much fatigue and suffering might have been spared by using trucks. Accordingly two were made by our carpenters at Fort Chipewyan, in 1827, for the return of the Expedition, and they answered extremely well. I mention this circumstance, in the hope that some such expedient will be adopted by the Traders for the relief of their voyagers, who have twice in every year to pass over this ridge of hills.

The annexed accurate drawing, taken by Lieutenant Back, from the highest part of this Portage, gives a beautiful delineation of one of the most picturesque scenes in the northern parts of America.

Tuesday,
12th.
Being now in a fair way to reach the Athabasca Lake, Dr. Richardson and I embarked, on the 12th, in the canoe, to proceed to Fort Chipewyan, for the purpose of preparing the gentleman in charge for the reception of the party.

By noon we got over the four Portages on Clear-Water river, and descended, with some trouble, the series of rapids that follow them. Once below these, the passage to the lake is generally considered as free from fatigue; but we did not find it so, for, owing to the shallowness of the water, the men had to get out

OF THE POLAR SEA.

and drag the canoe in several places. The difference between the depth of water now and in other years at the same period, was attributed to the snow having fallen in the preceding autumn before the frost was sufficiently intense to harden the ground, and, consequently, much of the moisture had penetrated the earth, which, under other circumstances, would have remained in a frozen state, for the supply of the river at the spring thaw.

In the course of the night we were under some alarm for one of our men, who having incautiously lain down to sleep under a wet sail, while the rain was pouring heavily, was seized with a cramp in the stomach, and violent pain in the head. Having been brought into the tent and covered with blankets, he became better before morning, but not sufficiently strong to allow of our setting off at the usual hour. We entered the Elk, or Athabasca River, at three P.M., on the 13th, and were carried swiftly down by its current to the Hudson-Bay Company's post named Berens House, where we stayed the night. Here we received a supply of dried meat.

We safely arrived in the Athabasca Lake on the 15th, by the channel of the " Riviere des Eaux remuées;" but in the subsequent traverse between Bustard Island and Fort Chipewyan the canoe was in danger of foundering in a sudden gale. Two large waves broke with full force into it, and obliged us to bear away and steer for the nearest shore; but the men having soon rested, and being now sheltered by islands, we pushed on to Fort Chipewyan. Our arrival there caused great surprise to its inmates, when they learned that we had come from England to that advanced post so early in the season, being only two days later than the time at which Dr. Richardson and Mr. Hood had arrived in 1819, though they had passed the winter at Cumberland House.

The stores at Fort Chipewyan being well furnished with warm

clothing, and other articles, which we required for the use of the men and Indians at winter-quarters, I availed myself of the permission which the chief factor of this department, Mr. James Keith, had given me to complete our stock of cloth, blankets, nets, and twine, to a quantity sufficient for two years' consumption. A supply of twine was indispensable, because, by a letter from Mr. Dease, I had learned that the meshes of the nets made in England, of the size generally required for fishing throughout this country, were too large for the smaller fish that frequent that part of Bear Lake where our house was to be constructed. Mr. Campbell, the clerk in charge, cheerfully gave me the benefit of his experience in making out lists of such things as we were likely to want, and in assorting and packing them.

Monday, 18th.

The boats rejoined on the 18th, and the crews were allowed the following day to recruit themselves. A party of Indians came very opportunely with fresh meat, which is always an agreeable change to the voyager, who has generally to live on dried provision. The Indians, as well as the women and children of the fort, spent the greater part of the day by the side of our boats, admiring their whole equipment, but more especially the gay figures painted on them. Many of these were different from any animals or representations they had seen, and, judging from the bursts of laughter, some curious remarks were made on them.

Wednesday, 20th.

It being necessary that I should await the arrival of Lieutenant Back's canoes, Dr. Richardson undertook to proceed with the boats towards Slave Lake. Their lading was now increased by the bales already mentioned, as well as by several bags of pemmican, which Mr. Keith had stored up for our use. The crews, however, were reduced by the discharge of three Englishmen, at their own desire, who thought themselves unequal to the fatigue of the service.

39

I had the happiness, on the 23rd, of welcoming my friends, Lieutenant Back and Mr. Kendall, on their arrival with three canoes. Their journey from Fort William had been expeditious, notwithstanding the detention of eighteen days, by bad weather, on the road. A serious misfortune had happened at the very outset of the journey, through the unskilfulness of one of the bowmen, in allowing his canoe to turn round and get before the current, while attempting to ascend the Barrier Rapid, by which it was driven against a stone with such force, as to be overset and broken. The stores were fortunately saved, though completely drenched; but many of the delicate atmospherical instruments were broken. Mr. Kendall was despatched to Fort William for another canoe while the things were drying.

On a subsequent occasion, in the Winipeg River, the same man placed his canoe in such a situation, as to endanger its being hurried down a steep fall, and had it not been for the coolness of a man, named Lavallé, who jumped into the water and held the canoe, while the rest of its crew arranged themselves so as to drag it into a place of safety, every life must have been sacrificed. The success, indeed the safety, of this kind of river navigation, among currents and rapids, depend on the skill of the bowman; and after these proofs of his incapacity, Lieutenant Back very properly engaged a substitute at the first fort to which he came.

At another time, in the Sturgeon-weir River, the canoe in which Mr. Kendall was embarked, having been accidentally driven before the current, she was only saved from destruction by his own powerful exertion and activity.

These short details will convey an idea of the anxiety and trouble these officers experienced in their journey to Chipewyan.

The party and the stores having now passed the more difficult part of the road, I discharged as many of the Canadians as could be spared, and furnished them with a canoe to take them home. Some went to Montreal; and they were the first persons who had ever gone from that place to Chipewyan, and returned in the same season.

Monday,
25th.

The greater part of the 25th was employed in obtaining astronomical observations, the results of which, we were delighted to find, placed Fort Chipewyan within a few seconds of longitude of the position in which it had been laid down on the former Expedition. Our present azimuth compasses showed an increase in variation, since 1820, of 2° 16′ E. The dip was observed 81° 26′ 47″.

Fort Chipewyan was this summer visited, for the first time, by a large flight of swallows, resembling the house-martins of England. They came in a body on the 25th of June, and immediately began to construct their earthy nests under the ledge of the south-front of the house. Some barn or forked-tail swallows also arrived on the 15th of June, and took possession of the store-houses and garrets, as they had in former years done. Some of the young of the last-mentioned birds were sporting on the tops of the houses as early as the 17th of July.

At sunset we embarked in four canoes, one having been procured here. The descent to Slave Lake occupied four days, and was unattended with any circumstance deserving mention, except that two of the canoes were broken in consequence of the guide mistaking the proper channel in a rapid; fortunately, these bark vessels are soon repaired, and we had only to regret the delay the accident occasioned.

We halted at the Salt River to take in salt, as we found, by a note left here, Dr. Richardson had done. The geese were moulting at this time, and unable to fly; they afforded us much sport in their chace, and an excellent supper every night.

1825.
July.

A body of Indians were waiting near the entrance of the lake to welcome our arrival; they were so numerous, that we were forced to omit our general custom of giving a small present to each native, and thus incurred the charge of stinginess, which the loud vociferations they raised on our setting sail, were probably meant to convey.

Friday,
29th.

At six, on the evening of the 29th, we reached Fort Resolution, the only establishment now at Slave Lake, and we felt happy in being once more under the roof of our hospitable friend, Mr. Robert M'Vicar, to whom I am much indebted for the excellent order in which he had brought up our supplies from Canada in the preceding year. Dr. Richardson, after a halt of two days, had gone forward with the boats.

All the portages on the road to Bear Lake being now passed, the Canadians made a request, that we would allow them to commemorate the event by a dance. It met with a ready compliance; and though they had been paddling for thirty-six out of the thirty-nine preceding hours, they kept up their favourite amusement until daylight, to the music of bagpipes, relieved occasionally by the Jews' harp.

We rejoiced to find at this post our worthy old Copper-Indian friends, Keskarrah and Humpy, the brother of Akaitcho, who had been waiting two months for the express purpose of seeing us. These excellent men showed that their gratification equalled ours, by repeatedly seizing our hands and pressing them against their hearts, and exclaiming, "How much we regret that we cannot tell what we feel for you here!" Akaitcho had left the fort about two months on a hunting excursion, hoping to return, with plenty of provision for our use, by the middle of August, which was as early as he thought we should arrive. Keskarrah confirmed the melancholy report we had heard in the more southern districts, that most of the hunters who had been in our service at Fort Enterprise, had been treacherously murdered, with many others of the tribe, by the Dog-ribs, with which nation we also learned the Copper-Indians had been at war, since the year of our departure from them, till the last spring. The peace had been effected through the mediation of Messrs. Dease and M'Vicar, and we were gratified to find that Akaitcho and his tribe had been principally induced to make this reconciliation, by a desire that no impediment might be placed in the way of our present expedition. " We have too much esteem," said Akaitcho, " for our father, and for the service in which he is about to be again engaged, to impede its success by our wars, and, therefore, they shall cease;" and on being asked by Mr. Dease whether he and some of his young men would go to hunt for the party at our winter quarters, he replied, " Our hearts will be with them, but we will not go to those parts where the bones of our murdered brethren lie, for fear our bad passions should be aroused at the sight of their graves, and that we should be tempted to renew the war by the recollection of the manner of their death. Let the Dog-ribs who live in the neighbourhood of Bear Lake furnish them with meat, though they are our enemies." Such sentiments would do honour to any state of civilization, and show that the most refined feelings may animate the most untutored people. Happily we were now so circumstanced as to be able to reward the friendship of these good men by allotting from our stores a liberal present to the principal persons. On the delivery of the articles to Keskarrah and Humpy, I desired them to communicate to Akaitcho, and the whole tribe, the necessity of their strictly adhering to the terms of peace, and assured them that I should not fail to urge the same obligations on the Dog-ribs. A silver royal medal, such as is given to the Indian chiefs in Upper Canada, was likewise left with Mr. M'Vicar, to be presented to Akaitcho, as a further mark of our regard for his former services and present good wishes.

The party was detained at Fort Resolution until this morning by a strong south-west gale; and even when we embarked, the wind and waves were still high, but time was too precious to allow of our waiting when there was a prospect of making any advance. As our future course inclined to the westward, we now quitted the track of the former journey to Fort Enterprise, along which we had been travelling from Lake Winipeg. We first steered for the Buffalo River, and then along the south shore of Slave Lake, obtained the latitude 61° 1′ N. at noon, and afterwards the longitude 114° 18½′ W. at the Isle of the Dead. The islands and shores of this part of the lake are composed of horizontal beds of limestone, containing pitch and shells.

A small party of Chipewyan Indians, with their principal chief, joined us at the encampment, from whom we learned that they had supplied Dr. Richardson with dried meat the preceding noon, at Hay River. The chief was very importunate for rum, but I steadily adhered to the determination I had formed this time, on my entering the Fur Country, of not giving spirits to any Indian. A share of our supper and tea, and some tobacco, were offered to him, and accepted, though with a bad grace. The Fur Company ceased the following season to bring any rum to this quarter, and I learned that this man was one of the few natives who were highly displeased at this judicious change.

We coasted this day along the low shore of the lake, steering from point to point to avoid the sinuosities of several deep bays, and passed the mouth of the Sandy and Hay Rivers, whose positions we settled by astronomical observations. The results of these, and the subsequent observations, appear in a

C 2

1825.
August.

Tuesday,
2nd.

table in the Appendix, and, therefore, will not be inserted, except at particular points.

On the 2nd we came to the narrow part between the Big Island of Mackenzie, and the main shore, and perceived that a gentle current was setting towards the Mackenzie River. The water in this strait is very shallow, and also in many places near the south shore, though we know, from trial, on the former Expedition, that the depth of the east end of the lake, at a distance from the land, exceeds sixty and seventy fathoms. The beach, both of the north and south shores of the strait, is strewed with drift timber. In clear weather the north shore is visible from the point of the south shore nearest Big Island.

Below this *detroit* the shores recede so as to form a small shallow lake, about twenty-four miles long, by from four to twelve miles broad, near the north-west end of which we encamped, in latitude 61° 15′ N., longitude 117° 6′ W. This spot may be considered as the commencement of Mackenzie River. The ground is very swampy, and nourishes willows only; but inland, at a short distance from the beach, grow plenty of the spruce-fir, poplar, aspen, and birch trees; and among the underwood, numerous shrubs and berry-bearing plants.

Wednesday,
3rd.

On the 3rd we travelled to another contraction of the river about one mile broad, through which the current sets between high banks with such force as to form strong eddies. There are likewise in this part many sandy islands, and through the channels between them the current rushed with no less rapidity than in that we descended. For distinction's sake, these islands have been named the " Isles of the Rapid:" below them occurs another expansion, which is called by the voyagers " The Little Lake;" and Sandy Point, at its northwest end, is considered by them as the commencement of the Mackenzie River.

When abreast of this point, a favouring breeze enabled us to use the sail as well as the paddles, and with the assistance of the current great progress was made. We had occasional glimpses of the Horn and Rein-Deer Mountains as we passed along; but, until we were some some way below the rapids, our view was very limited, owing to the woods being on fire in almost every direction. This I should have mentioned to have been the case in many parts between Isle à la Crosse and the Mackenzie. The cause of these extensive conflagations I could not learn; some attributed them to voluntary acts of the Indians, and others to their negligence in leaving their fires burning.

We put up at sunset on a beach of gravel under a well-wooded bank of moderate height, and the party regaled themselves with raspberries and other indigenous fruits.

At half past two A.M., on the 4th, the canoes were again on the water, and being driven by sail and current, made good way. We stopped at the Trout River, which flows in from the southward, and ascertained its longitude 119° 47′ W. The breadth of the Mackenzie is here about two miles, and its banks are composed of a muddy clay: the stones on the beach mostly limestone, with some boulders of primitive rocks. The trees are of the kinds we had seen north of the Athabasca Lake; they are here of a smaller size. Five miles below this part, the Mackenzie is divided into several channels by islands, and the current runs with increased swiftness, and strong eddies.

The latitude 61° 26′ 30″ N. was obtained at noon; it was the same as on the preceding day; so that our course, in the interval, had been due west.

The banks now were higher, and for the next forty miles the breadth of the stream did not exceed one mile, nor was less than half a mile; its course inclined more to the north. We passed

1825.
August.

the site of the first establishment that the North-West Company had made in these parts, which was erected by Mr. Livingstone, one of the partners, who, with the whole of the crew of his canoe, except one individual, were massacred by the Esquimaux on the first attempt to open a trade with them.

At three P.M. a picturesque view opened upon us of a distant range of mountains running east and west, and nearly at right angles to the course of the river. The current being considerably increased by the contribution of some streams near this place, we descended very swiftly. Six miles below Pine Island, there is a strong but not a dangerous rapid; and about fifteen miles farther is Fort Simpson, the principal depôt of the Hudson-Bay Company for this department, at which we arrived by eight P.M., and thus escaped a very wet, comfortless night. Dr. Richardson had departed for Fort Norman the preceding day.

This establishment, three hundred and thirty-eight miles from Fort Resolution on Slave Lake, is situated at the confluence of the River of the Mountains and the Mackenzie. The former is the channel of communication with a fur post not far distant from the Rocky Mountain Range, from whence the residents here procure much of their provision, including a tolerable supply of potatoes, which have been recently introduced from the southern parts. Mr. Smith, the chief factor of the district, was fortunately at Fort Simpson, so that I had the opportunity of arranging with him as to supplies of provision or stores that my party might require during its residence at Bear Lake. He cheerfully acceded to every suggestion that was made, and likewise furnished me with a letter of instruction to the same effect, addressed to the gentlemen in charge of the lower posts.

I learned from Mr. Smith that, as yet, a few only of the Indians who live nearest the mouth of the river, and none of the Esquimaux, had been apprized of our intended visit, the traders at the lower posts having considered that it would be better to defer this communication until we should arrive in the river, for fear of disappointing these people, which might have been attended with unpleasant results.

There were two Canadians here belonging to the Expedition, whom Mr. Dease had sent to serve as guides to Bear Lake. By letters which they brought, I was informed that Indian hunters were engaged, and the necessary buildings in course of preparation for our reception. As Fort Simpson had been short of ammunition during the summer from some accidental cause, I was glad to find that Mr. Dease had been enabled to lend from our stores a barrel of powder, and a bag of balls, and I now increased the loan, so as to meet the probable demands of the Indians, until the Company's supplies should arrive, when they would return to Fort Norman the whole of what we had lent. Cloudy weather limited our astronomical observations at this place to the dip of the needle, which was observed 81° 54'.

We quitted the fort on the 5th, soon after noon, whence the river preserving nearly a straight course for fifteen miles, gradually extends itself to nearly two miles in breadth: in its channel there are three islands. At two P.M. we obtained the first glimpse of the Rocky Mountains, and kept them in view until we encamped, which was early, as the canoes required gumming. The outline of the mountains was very peaked, and at their easternmost part was a cone-shaped hill, higher than the rest, whose summit was veiled by clouds. The general appearance of the range somewhat reminded me of the east end of Jamaica.

The morning of the 6th was beautifully fine: we embarked at 2ʰ 30' A.M., and by seven came within six or seven miles of the mountain range, where the river suddenly changes its course from W.b.N. to north, in longitude 123° 31′ W.

A distinct stratification was perceptible on the face of the nearest mountain : on one side of a nearly perpendicular ravine the strata dip to the southward at an angle of 25°; whilst on the other they are nearly horizontal. There was a large accumulation of debris at its base : every part of the hill was destitute of vegetation. Its altitude was guessed at one thousand two hundred feet.

At noon, in latitude 62° 49′ N., we saw a chain of mountains, on the eastern side of the river, similar in their outline and general character to those hitherto seen only on the opposite bank. Between these ranges the river flowed in a channel two miles broad; but as we advanced we receded from those on the western side, their direction being W.N.W. In the brilliancy of the sunshine, the surfaces of some of the eastern hills, which were entirely bare, appeared white as marble, and for some time we fancied them to be covered with snow. By four P.M. we reached the Rocky Island mentioned by Mackenzie, where, from the river being contracted, the current flowed with great rapidity, and soon brought us opposite to the remarkable hill close by the river side, which that persevering traveller ascended in July, 1789. His account, and the annexed detailed engraving, from an accurate drawing by Mr. Kendall, render a description of it unnecessary. It is composed of limestone, and is about four hundred feet high.

We continued a N.b.W. course for eight miles, and encamped at sunset, having travelled this day one hundred and twenty miles. A small supply of fresh deer's meat was obtained from some Dog-Rib Indians. Their canoes were made of the bark of the pine-tree, sewn at the ends and top with the fibrous parts of the root of that tree, leaving only a space sufficient for the legs of the sitter.

We pursued our course at dawn of day, and at the end of a few miles came to a more winding part of the river, where the stream is interrupted by numerous sand banks and shoals, which we had some trouble to get round. Mr. Kendall, in his Journal, remarks of this part, " That bubbles of air continually rose to the surface with a hissing noise resembling the effervescence produced by pouring water on quick lime."

We arrived at Fort Norman at ten, A.M., distant two hundred and thirty-six miles from Fort Simpson, and five hundred and seventy-four from Fort Resolution.

Being now only four days' journey from Bear Lake, and there remaining yet five or six weeks of open season, I resolved on following up a plan of a voyage to the sea, which I had cherished ever since leaving England, without imparting it to my companions, until our departure from Fort Chipewyan, because I was apprehensive that some unforeseen accident might occur in the course of the very intricate and dangerous river navigation between Fort William and the Athabasca Lake, which might delay our arrival here to too late a period of the year. It was arranged, *first*, that I should go down to the sea, accompanied by Mr. Kendall, and collect whatever information could be obtained, either from actual observation, or from the intelligence of the Loucheux Indians, or the Esquimaux, respecting the general state of the ice in the summer and autumn ; the direction of the coast, east and west of the Mackenzie ; and whether we might calculate upon any supply of provision. *Secondly*, Dr. Richardson, on his own suggestion, was to proceed in a boat along the northern shore of Bear Lake, to the part where it approached nearest to the Coppermine River, and there fix upon a spot to which he might bring the party the following year, on its return from the mouth of that river. And, *thirdly*, that these undertakings might not interfere with the important operations necessary for the comfortable residence and subsistence of the Expedition during the following winter, Lieutenant Back was to superintend them during my absence, with the assistance of Mr. Dease, chief trader of the Hudson Bay Company, whose suggestions, relative to the proper distribution of the Indian hunters, and the station of the fishermen, he was to follow. Accordingly, Dr. Richardson, on his quitting this place two days previous to our arrival, had left the largest of the boats, the Lion, for my use, and a well-selected crew of six Englishmen, and Augustus the Esquimaux.

Lieutenant Back was directed to take the canoes forward to Bear Lake, laden with such supplies as would be required for the winter, and was further instructed to furnish Dr. Richardson with one of the boats, and a good crew. The services of the Canadians who had brought the canoes from Penetanguishene, being no longer required, I desired Lieutenant Back to discharge them, and also all the voyagers of Mr. Dease's party who could be spared. They were sent in canoes to Slave Lake, where I had arranged with Mr. M'Vicar for their being supplied with the means of gaining subsistence by fishing, during the winter; and the following spring, they were to be forwarded to Canada, at the expense of Government, according to the terms of their agreement.

Fort Norman being situated in our way to the sea, the pemmican and other stores, intended for the voyage along the coast next season, were deposited here, by permission of Mr. Smith, under the care of Mr. Brisbois, the clerk in charge. Our observations place this establishment in latitude 64° 40′ 30″ N., and longitude 124° 53′ 22″ W.

The above matters being satisfactorily settled, and a few articles packed up as presents to the Indians and Esquimaux,

1825.
August.

Mr. Kendall and I embarked on the 8th, at noon, taking, in addition to our crew, a voyager, who was reported to be able to guide us through the proper channels to Fort Good Hope, of which, however, we found him altogether ignorant. We were accompanied by Lieutenant Back, with the three canoes, each manned by five men. The crews of the canoes imagining they could easily pass our English boat, were much surprised, on putting it to the proof, to find the boat take and maintain the lead, both under sail and with oars.

A few miles above the Bear Lake River, and near its mouth, the banks of the Mackenzie contain much wood coal, which was on fire at the time we passed, as it had been observed to be by Mackenzie in his voyage to the sea. Its smell was very disagreeable. On a subsequent trial of this coal at our winter quarters, we found that it emitted little heat, and was unfit for the blacksmith's use. The banks likewise contain layers of a kind of unctuous mud, similar, perhaps, to that found on the borders of the Orinoco, which the Indians, in this neighbourhood, use occasionally as food during seasons of famine, and even, at other times, chew as an amusement. It has a milky taste, and the flavour is not disagreeable. We used it for whitening the walls of our dwellings; for which purpose it is well-adapted.

The entrance of the Bear Lake River is distinguished by a very remarkable mountain, whose summit displays a variety of insulated peaks, crowded in the most irregular manner. It is composed of limestone; and from the lower cliffs, which front the river, a dark, bituminous liquid oozes and discolours the rock. There are likewise two streams of sulphureous water that flow from its base into the Mackenzie. At this place we parted from our friend, Lieutenant Back, who entered the clear and beautiful stream that flows from Bear Lake, of whose pure waters we had also the benefit, till they were overpowered by the muddy current of the Mackenzie. The day was fine, the wind fair, the current swift, and every circumstance concurred to put the party in high glee. There was little in the scenery to attract our attention, now that we had become familiar with the general appearance of the Mackenzie, and we passed island after island, of the same alluvial mud, without further regard than the delineation of them in the survey book. At length, however, a most picturesque view of the rocky mountain range opened before us, and excited general admiration, and we had also some portions of the mountain range on the eastern side of the river, in view for the remainder of the day's journey. The outline of these mountains is very irregular, the highest parts being peaked hills. The general direction of the ranges is between N.W. and N.W.b.W.

Being unwilling to lose the advantage of the wind, we only put ashore to sup, and after two hours' delay, resumed our voyage under easy sail. When the sun rose, the oars were used; and then, as the current set at the rate of two miles and a half per hour, the boat travelled swiftly down the stream. The eastern bank of the river, along which we were passing, is about one hundred and twenty feet high, almost perpendicular, and is composed of thin strata of bituminous shale. Amongst the fragments of shale which strewed the beach, we found many pieces of brown wood-coal. A reach, eighteen miles in length,

followed. It is bounded on both sides by high cliffs of sandstone. We landed to breakfast, and to obtain the longitude, 128° 23′ W.

From the reach here described, are seen two hills, named by me the East and West Mountains of the rapid, which seem to present a barrier to the further progress of the stream; but the river, bending suddenly between them to the north, dilates into a kind of basin, and, by so doing, opens by far the most interesting view of the Rocky Mountains which the Mackenzie affords. The river, too, makes its nearest approach to those mountains at this spot, and, probably, the easiest communication with them would be by ascending a small stream that flows in here on the western side. Here too are found the first rapids mentioned by Mackenzie, which continue in succession for two miles, when the water is low. The centre of the basin is occupied by low sandy islands; and the channel on the western side is the deepest. The beauty of this scene furnished employment for the able pencil of Lieutenant Back, on a subsequent occasion. As the Mackenzie, in its further descent, continues to hold a northerly course, and the range of mountains runs N.W.b.N., we did not obtain any other view of them till we approached the sea.

At one P.M. we saw a party of Indians encamped on the beach of a small stream, whom we invited to come off to us. They hesitated at first, being doubtful who we were, from our boat being different in shape from any they had seen, and carrying two sails; but after some time they launched their canoes, and brought us a good supply of fresh deer's meat. The sight of our boats seemed to delight them as much as the ammunition and tobacco which they received. These were Hare Indians, the tribe that follows next to the Dog-Ribs, in the line of country below Bear Lake; and, like them, they speak a dialect of the Chipewyan language. We admired the shape and appearance of their canoes, which were larger than those used by the Chipewyans, and had the fore part covered with bark, to fit them for the navigation of this broad river, where the waves are often high.

The river varied from two to four miles in breadth, and its course was interrupted by several small islands and sand-banks. At six P.M. we came to an open space, bounded by lofty walls of sandstone.   In this expansion are found the second rapids of Mackenzie : at the first appearance they seem dangerous, but are not so.   The river becomes again contracted, and rushes with great force for the space of seven miles through a kind of defile, varying in breadth from four hundred to eight hundred yards, which has been appropriately named " The Ramparts," by the traders.   The walls of this defile are from eighty to one hundred and fifty feet high, and are composed of limestone, containing numerous shells : for a part of the way the stone is very white, and in the rest it is blue.   Several streams of water were running over the summits of the cliffs, which had worn the stone, in some places, into a turreted shape; while the heaps, overthrown by its action at their base, resembled mounds for defence. To these appearances were occasionally added cavernous openings, and other hollow parts, not unlike the arched windows or gateways of a castellated building.   I could not help fancying what delight a visit to this spot would afford to any person of a romantic turn, especially at the time we first saw it, when the broad shadows of a declining sun gave effect to the picture. This is a place of resort for the Hare Indians to fish, and we were visited by a large party of men and women of that tribe, who brought fish, berries, and meat.   They were all neatly clothed in new leathern dresses, highly ornamented with beads and porcupine quills.   The paintings of animals on the sides of our boats were very attractive to them; they scanned every figure over and over, bursting into laughter whenever they recognised any of the animals.   We encamped near a small river below the ramparts, one hundred and ninety-three miles from Fort Norman.   Two young Indians followed us in their canoes, bringing some musk rat-skins, and fish, for sale.   We purchased the fish, but declined taking the furs.   They were so pleased with their reception, that they passed the night by our fire.

At day-light we again embarked, and descended the river pleasantly and swiftly under sail, having the benefit of a strong current, especially where it was narrowed by islands or sand-banks. The sides of the river are generally high cliffs of limestone or sandstone, and its breadth from two to three miles. The intervals between these cliffs are mostly occupied by hills of sand, from eighty to one hundred feet high, whose intermediate valleys are well wooded; and whenever these occur, the channel of the river is much interrupted by banks, on which, as well as on the beach, there are vast collections of drift timber, piled, in some places, twenty feet high, by the spring floods.

At eleven P.M. we arrived at Fort Good Hope, the lowest of the Company's establishments; it is distant from Fort Norman three hundred and twelve miles, and is in latitude 67° 28' 21" N., and longitude 130° 51' 38" W.: the variation of the compass being 47° 28' 41" E. Our arrival at this period of the year, at least two months earlier than that of the Company's boats from York Factory, caused great astonishment to the few inmates of this dreary dwelling, and particularly to its master, Mr. Charles Dease, who scarcely recovered from his surprise until we had been seated some time in his room. But this over, he quickly put every one in motion to prepare a meal for us, of which we stood in much need, as it was then verging on midnight, and we had breakfasted at eight in the morning. This post had been but recently established for the convenience of the tribe of Indians whom Mackenzie calls the Quarrellers, but whom the traders throughout the fur country name the Loucheux. As this name is now in general use, I shall adopt it, though it is. but justice to the people to say that they have bright sparkling eyes, without the least tendency to that obliquity which might be inferred from the term. The fact is, that Loucheux, or Squinter, was intended to convey the sense of the Indian name

1825.
August.
of the tribe—Deguthée Dennee, which means " the people who avoid the arrows of their enemies, by keeping a look out on both sides." None of the tribe was at this time at the fort; but from Mr. Dease we learned the interesting fact, that the Loucheux and Esquimaux, who are generally at war, had met amicably the preceding spring, and that they were now at peace. We procured from the store an assortment of beads, and such things as were most in request with the Loucheux, and made up a small package of clothing to be presented to each chief of that tribe, whose favour it was thought advisable by this means to propitiate, as they were the next neighbours to the Esquimaux.

Thursday,
11th.
After the latitude had been observed, we embarked, and were accompanied by Mr. Dease as far as Trading River, where he expected there might still be a party of Indians, which did not prove the case. This river being the usual limit of the traders' travels towards the sea, the voyager who had come with us from Fort Norman declined going any farther, and by permission of Mr. Dease he was exchanged for a young half-breed named Baptiste, the interpreter of the fort, who went under the promise of being left with the chief of the Loucheux, to whom he was to introduce the party.

The reach below Trading River is remarkable, from the banks on the eastern side consisting of hills of a light yellow marl-slate, nearly uniform in shape, and strongly resembling piles of cannon shot. The name of Cannon-Shot Reach was, therefore, bestowed on it. The channel of the river is very intricate, winding amongst numerous sand-banks, and some low alluvial islands, on which willows only grow. Its breadth is about two miles, and the depth of water, in the autumn, from six to twelve feet. In passing through Cannon-Shot Reach, we were hailed by an Indian from the shore, and landed immediately, to inform him of the purport of our visit. As soon as Baptiste had

explained these matters to him, the man, deeming it of importance that we should be properly introduced to his relatives,
offered to accompany us to the next party, providing we would
undertake to carry his baggage. This we consented to do, little
expecting, from the appearance of poverty in himself and his
family, and still less from that of his tent, a mere covering of
bark and pine branches, supported on three poles, that load upon
load of unsavoury fish would be tossed into the boat. However,
we were unwilling to retract our promise, and suffered our vessel
to be completely lumbered. We then pushed off, leaving the
family to follow in the canoe, but in a short time our ears were
assailed by the loud cries of the man demanding that we should
stop. On his coming up, we found he was apprehensive of the
canoe sinking, it being very leaky and overloaded, and of his
losing his wife and infant child. The water being thrown out,
the man proposed going forward and keeping by our side.
There was nothing now to fear, yet the lamentations of the
woman became louder and louder, and at last the poor creature threw off her only covering, raised the most piteous
cries, and appeared a perfect object of despair. We learned
from Baptiste that she was mourning the loss of two near relatives who had recently died near the spot we were passing. In
this manner do these simple people show their sorrow for the
death of their connexions. As we drew near the tents of the
party on shore, the husband proclaimed with a stentorian voice
who we were; this produced a long reply, of which Baptiste
could only collect enough to inform us that many persons were
lying sick in the lodges, and that two had died the preceding
day. Not choosing to expose ourselves to the hazard of con-
tagion, we put the baggage of our friend on shore at some
distance below the lodges. All those who were able to manage
a canoe, came off to receive presents, and to see Augustus, the

1825.
August.

principal object of attraction. Each person crowded to the side on which he sat to shake him by the hand; and two of the party, who had been occasionally with the Esquimaux, contrived to make him understand that, being accompanied by him, we need apprehend no violence from them, though they were a treacherous people. At the end of five miles farther we put on shore to sup, and afterwards slept in the boat; but Augustus spread his blankets on the beach before the fire, and allowed four of the Loucheux, who had followed us from the tents, to share them with him.

Friday, 12th.

At daylight we loosened from the beach, and continued with the descent of the river; winding, in our course, as numerous sandbanks rendered necessary. In a few hours we descried another collection of Indian lodges. One of the party happened to be examining his nets nearer to us than the tents; on espying the boat, he immediately desisted, and paddled towards his friends with the utmost speed, bawling the whole way for them to arm. The women and children were seen hurrying up the bank to hide themselves; and by the time we had got abreast of the lodges, the whole party were in a state of defence. They stood on the beach gazing at us evidently with much distrust; and for some time no one would accept our invitations to approach. At length an adventurous youth, distinguishable among the rest by the gaiety of his dress, and the quantity of beads that were suspended around his neck, launched his canoe and paddled gently towards the boat, till he discovered Augustus, whom he knew by his countenance to be an Esquimaux; then rising from his seat, he threw up his hands for joy, and desired every one of the party to embark at once. The summons was instantly obeyed, and a friendly intercourse followed; each person that had a gun discharging its contents, and taking the iron heads and barbs from the arrows, to show their entire confidence. On landing to breakfast, we found that the dialect of this party was different from that of the men we had seen yesterday, and that Baptiste did not understand their language; consequently our communications were carried on by signs, except when they attempted to speak Esquimaux, which Augustus, with difficulty, made out. He was still the centre of attraction, notwithstanding Mr. Kendall and myself were dressed in uniform, and were distributing presents to them. They caressed Augustus, danced and played around him, to testify their joy at his appearance among them, and we could not help admiring the demeanour of our excellent little companion under such unusual and extravagant marks of attention. He received every burst of applause, every shake of the hand, with modesty and affability, but would not allow them to interrupt him in the preparation of our breakfast, a task which he always delighted to perform. As soon as we had finished our meal, he made his friends sit down, and distributed to each person a portion of his own, but without any affectation of superiority. When we were on the point of embarking, the oldest Indian of the party intimated his desire that we should stop until some one whom he sent for should come. This proved to be his son, in a very sickly state. Though the day was warm, the lad was shivering with cold, and it was evident he was suffering from fever, which the father had no doubt we could cure. The only remedy we could apply was some warm tea, with a little brandy in it, which we afterwards learned had the desired effect of restoring the invalid. Again we were preparing to set off, when the same old man begged us to stop until the women should come; these were no less pleased with Augustus, and with the presents they received, than the men had been.

This good-natured tribe is distinguished by the traders as the Lower Loucheux, but the literal meaning of their Indian name is the Sharp Eyes. They are decidedly a well-looking people:

in manner, and general appearance, they resemble the Esquimaux near the mouth of the Mackenzie, though not in their eyes, which are prominent and full.   Their canoes, too, are shaped like those of the Esquimaux, and made of birch bark, which, by some process, is striped from the gunwale perpendicularly downwards, for the purpose of ornament.   Their summer dress, like that of the upper Loucheux and Esquimaux, is a jacket of leather, prolonged to a point before and behind : the leggings, of the same material, are sewn to the shoes, and tied by a string round the waist.   The outer edges of their dress are cut into fringes, coloured with red and yellow earth, and generally decorated with beads.   Beads are so much coveted by them, that, for some years, they were the principal article of trade exchanged for their furs; and even now the successful hunter, or the favourite son, may be known by the quantity of strings of different coloured beads which he has about his neck.   These Indians are the only natives of America, except the Esquimaux, whom I have seen with the septum of the nose perforated, through which, like the Esquimaux, they thrust pieces of bone, or small strings of shells, which they purchase from that people.   Few of them have guns, but each man is armed with a bow and arrows.   The bows are constructed of three pieces of wood, the middle one straight, and those at each end crooked, and bound with sinews, of which the string is also made.   The dress of the women only differs from that of the men by the hood being made sufficiently wide to admit of their carrying a child on their back.

At ten A.M. we resumed our journey, followed by the young man who had first spoken to us, and his brother, in their canoes, and in the course of two hours came abreast of a remarkable round-backed hill, on which we were informed Mr. Livingstone and his party had encamped in 1795, the night before they were massacred.   This hill marks the commencement of another contraction of the river, which is here pent in between very steep cliffs of blue limestone, which I have denominated the Narrows. The Red River contributes its waters to the Mackenzie at the lower part of the Narrows, in latitude 67° 27′ N., longitude 133° 31′ W.; and, though of inconsiderable size, is remarkable as being the boundary between the lands claimed by the Loucheux Indians and those of the Esquimaux, and likewise as the spot where the amicable meeting between these tribes had been held in the preceding spring. We did not find the chief of the Loucheux here, as had been expected, and therefore passed on. The banks of the river, now entirely composed of sand and sandstone, became gradually lower, and more bare of trees. At the end of eight miles we arrived at a very spacious opening, in which were numerous well-wooded islands, and various channels. The rocky mountains on the west once more appeared in view, extending from S.W. to N.W., and preserving a N.W.½W. direction; and of this range a very lofty peak, and a table mountain, which I have named after the late Mr. Gifford, form the most conspicuous features. We steered into the eastern channel, as being that through which the current seemed to run swiftest; and as soon as we came to a high bank we landed, for the purpose of taking a survey of the surrounding scene. But even from its summit our view was very limited, and all we could discover was, that we were certainly in that expansion of the river that Mackenzie delineates in his chart, and, therefore, in the fair way to the sea, whatever channel we took. This might have been inferred, from the sudden departure of our two Indian companions, who dropped behind and turned their canoes round, without further ceremony, as soon as they saw our intention of entering the eastern channel. Baptiste, who was asleep at the time, expressed surprise at their having gone back, but consoled himself with the idea of meeting the Indian chief the next morning, at a place he called the Forks.

We were amused at conjecturing how great his surprise would be should he be next disturbed by the hallooing of a party of Esquimaux, whom he greatly dreaded. At the end of twenty-three miles descent in the middle channel, having passed one that branched off to the eastward, we put up at an early hour, and caused the guns to be cleaned, and two sentinels appointed to watch, lest the Esquimaux should come upon us unawares. The banks of the river, as well as the islands, are entirely alluvial, and support willows at the lower parts, and the spruce-fir trees at the summits. The beach on which we were encamped was much intersected with the recent tracks of the moose and rein-deer.

We embarked at three A.M. on the 13th; and as we were in momentary expectation of meeting the Esquimaux with whom I wished to have an interview, the masts were struck, lest they should discover the boat at a distance, and run off. We soon passed two of their huts, which did not seem to have been recently inhabited. The longitude 134° 20′ 30″ W., and variation 51° 4′ 20″ E., were observed at the time we halted to breakfast, and the latitude 68° 15′ 50″ N., at noon. The Rein-deer mountains on the eastern side, came in view before noon. The range on the west was also occasionally visible: we were descending between the M'Gillivray and Simpson islands, in a channel that did not exceed half a mile in breadth. A fine breeze sprung up after noon, of which we took advantage by setting the sails, not having seen any recent traces of the Esquimaux. At the extremity of Simpson island there is a broad channel, which pours its waters into the one in which we were, at a place where the stream is contracted by a small island, and a strong rapid is the consequence of this junction. Here we found many huts, and other indications of its being a place of resort for fishing; here, too, it is supposed Mr. Livingstone and his crew fell a sacrifice to the first party of Esquimaux whom they met. Several other openings branched off to the eastward; but we continued to follow the largest channel, in which the current was very strong, and kept nearly parallel to, and about ten miles from, the Rein-deer mountains. Their outline, viewed from this distance, appeared very regular, the only remarkable parts being some eminences that were tinged with a deep pink colour. Sailing by one of the huts at a quick rate, every one's attention was arrested at hearing a shrill sound, which was supposed to be a human voice; but on landing to ascertain the fact, we could find no person, nor any footsteps. We, therefore, continued our journey. As we proceeded, the river became more devious in its course, the huts of the Esquimaux were now more frequent; none of them, however, seemed to have been recently inhabited. The islands were of the same alluvial kind as those seen yesterday, and the wood on them equally plentiful and large. We stopped to sup at nine, extinguished the fire as soon as we had finished, and then retired to sleep in the boat, keeping two men on guard.

We set off aided by a fresh breeze this morning, and at the end of seven miles came to the last of the fir trees, in latitude 68° 40′ N., the only wood beyond this being stunted willows, which became still more dwarfish at thirty miles from the mouth of the river. There was plenty of drift-wood on the borders of the islands, and some even on the higher parts, at a distance from the water; from which it would appear that at certain seasons they are inundated. At length the main stream took a turn to the S.S.W., which we followed, though there was a branch going northwards, but it seemed to be much impeded by mud-banks*. At the end of eight miles the river again inclined to the north of west, round the southern extremity of Halkett island, and there were openings to the north

* An attentive perusal of Sir Alexander Mackenzie's Narrative leads me to the conclusion, that it was this northern branch which that traveller pursued in his voyage to Whale Island.

and south, which we did not stop to examine. A fog-bank hung over the northern horizon, which gave us no little un-easiness, from its strong resemblance to a continuous line of ice-blink ; and the clouds, from the sun-beams falling on them, had the exact appearance of icebergs. However, the sun became sufficiently powerful in the afternoon to dissi-pate the cause of this illusion, and relieve us from anxiety on that score. A body of water, nearly equal to that we were descending, poured in between the Colville and Halkett islands with such force as to cause a very strong ripple at the point of junction, which we avoided by keeping close to the shore of Langley island. The channel, after the union of these streams, increased to a breadth of two miles, preserving a N.N.W. course. We stood twelve miles in this direction, and two to the west-ward, when we were gratified by the delightful prospect of the shore suddenly diverging, and a wide space of open water to the northward, which we doubted not would prove to be the sea. Just at this time a seal made its appearance, and sported about the boat as if in confirmation of this opinion. We attempted to coast along the shore of Ellice island, but found the water too shallow, and that the boat grounded whenever we got out of the channel of the river, which was near the western side. The wind and waves were too high for us to make any progress in the middle of the stream, and as the clouds threatened more bois-terous weather, we went to Pitt island to encamp. The haze which had hidden all distant objects since five P.M. passed off as the sun set, and we gained a very magnificent view of that portion of the rocky mountain which I have called after my companion Dr. Richardson, and of which the remarkable conical peak, named in honour of my friend Dr. Fitton, President of the Geological Society, and the Cupola mountain, are the most conspicuous objects. These were subsequently found to be near sixty miles distant. The[65] water was entirely fresh, and there was no perceptible rise of tide. Our drowsy companion Baptiste, when he looked upon the vast expanse of water, for the first time, expressed some apprehension that we had passed the Forks, and that there was a doubt of our seeing the Indian chief; but he was by no means convinced of the fact until the following day, when he tasted salt water, and lost sight of the main shore. After our Sunday evening's supper, the party assembled in the tent to read prayers, and return thanks to the Almighty, for having thus far crowned our labours with success.

In the morning of the 15th the wind blew a gale, as it had done through the night, and every object was obscured by a thick fog. About six A.M. we took advantage of a temporary abatement of the wind to cross over to some higher land on the eastern side, which we had seen the preceding evening, appearing like islands. Owing to the thickness of the fog, we were guided in our course at starting solely by the compass. When we reached the channel of the river, the gale returned with increased violence, and its direction being opposite to the current, such high waves were raised, that the boat took in a good deal of water. The fog now cleared away, and the three eminences mistaken for islands were ascertained to be conical hummocks, rising above the low eastern shore. We pushed for the nearest, and landed a short distance from its base at eight A.M. On going to the summit of this eminence, in the expectation of obtaining the bearings of several distant points, we were a little disappointed to find that only the low shores of Pitt Island were visible, extending from S.E. to W.N.W., though we were repaid for our visit by observing two moose deer quietly browsing on the tops of the willows, a short distance from us. Mr. Kendall hastened down to despatch Baptiste in pursuit of them, who returned an hour afterwards to inform us that he had wounded one, which he had been prevented from following by the loss of his powder-horn. As there was no possibility of our getting forward until the gale abated, Baptiste and Augustus were sent out to hunt, there being numerous tracks of moose and rein deer in the neighbourhood of the tent. I also despatched Mr. Kendall, with two seamen, to walk some distance into the interior, and endeavour to clear up the doubt whether we were upon the main shore, or upon an island. The astronomical observations obtained at the encampment place it in latitude 69° 3′ 45″ N., longitude 135° 44′ 57″ W. A tide-pole was put up immediately on our landing, and we perceived the water to rise about three inches in the course of the forenoon, and to fall the same quantity in the evening. The temperature of the air did not exceed forty-eight degrees all this day: when in the river, it used to vary from 55° to 70°. Mr. Kendall came back in the evening, bringing the agreeable intelligence that he had assisted in killing a female moose and her calf, and that Augustus had shot a reindeer. Some men were sent to carry the meat to the borders of a river which Mr. Kendall had discovered, while the boat went round to its entrance about one mile from the encampment. They returned at sunset. Many geese and ducks were seen by our hunters. Throughout the whole of Mr. Kendall's walk, of twelve or fourteen miles, he saw only the same kind of flat land, covered with the dwarf willow and the moose-berry plant, as was discovered from the tent, except one small lake, and the river that has been mentioned, issuing from it.

Tuesday, 16th.
The atmosphere was so thick on the morning of the 16th as to confine our view to a few yards; we therefore remained at the encampment till the sun had sufficient power to remove the fog: temperature of the air 39°. Embarking at eleven A.M., we continued our course along the shore of Ellice Island, until we found its coast trending southward of east. There we landed, and

were rejoiced at the sea-like appearance to the northward. This point is in latitude 69° 14′ N., longitude 135° 57′ W., and forms the north-eastern entrance to the main channel of the Mackenzie River, which, from Slave Lake to this point, is one thousand and forty-five miles, according to our survey. An island was now discovered to the N.E., looking blue from its distance, towards which the boat was immediately directed. The water, which for the last eight miles had been very shallow, became gradually deeper, and of a more green colour, though still fresh, even when we had entirely lost sight of the eastern land. In the middle of the traverse, we were caught by a strong contrary wind, against which our crews cheerfully contended for five hours, though drenched by the spray, and even by the waves, which came into the boat. Unwilling to return without attaining the object of our search, when the strength of the rowers was nearly exhausted, as a last resource, the sails were set double-reefed, and our excellent boat mounted over the waves in the most buoyant manner. An opportune alteration of the wind enabled us, in the course of another hour, to fetch into smoother water, under the shelter of the island. We then pulled across a line of strong ripple which marked the termination of the fresh water, that on the seaward side being brackish; and in the further progress of three miles to the island, we had the indescribable pleasure of finding the water decidedly salt.

The sun was setting as the boat touched the beach, and we hastened to the most elevated part of the island, about two hundred and fifty feet high, to look around; and never was a prospect more gratifying than that which lay open to us. The Rocky Mountains were seen from S.W. to W.½N.; and from the latter point, round by the north, the sea appeared in all its majesty, entirely free from ice, and without any visible obstruction to its navigation. Many seals, and black and white whales were sporting on its waves; and the whole scene was calculated to excite in our minds the most flattering expectations as to our own success, and that of our friends in the Hecla and the Fury. There were two groups of islands at no great distance; to the one bearing south-east I had the pleasure of affixing the name of my excellent friend and companion Mr. Kendall, and to that bearing north-east the name of Pelly was given, as a tribute justly due to the Governor of the Hudson-Bay Company, for his earnest endeavours to promote the progress and welfare of the Expedition. A similar feeling towards my much-esteemed friend Mr. Garry, the Deputy Governor of the Company, prompted me to appropriate his name to the island on which we stood,—a poor, indeed, but heartfelt expression of gratitude, for all his active kindness and indefatigable attention to the comfort of myself and my companions.

During our absence the men had pitched the tent on the beach, and I caused the silk union-flag to be hoisted, which my deeply-lamented wife had made and presented to me, as a parting gift, under the express injunction that it was not to be unfurled before the Expedition reached the sea. I will not attempt to describe my emotions as it expanded to the breeze— however natural, and, for the moment, irresistible, I felt that it was my duty to suppress them, and that I had no right, by an indulgence of my own sorrows, to cloud the animated countenances of my companions. Joining, therefore, with the best grace that I could command, in the general excitement, I endeavoured to return, with corresponding cheerfulness, their warm congratulations on having thus planted the British flag on this remote island of the Polar Sea.

Some spirits, which had been saved for the occasion, were issued to the men; and with three fervent cheers they drank to the health of our beloved monarch, and to the continued success of our enterprise. Mr. Kendall and I had also reserved a little of our brandy, in order to celebrate this interesting event; but Baptiste, in his delight of beholding the sea, had set before us some salt water, which having been mixed with the brandy before the mistake was discovered, we were reluctantly obliged to forego the intended draught, and to use it in the more classical form of a libation poured on the ground.

Baptiste, on discovering that he had actually reached the ocean, stuck his feathers in his hat, and exultingly exclaimed, " Now that I am one of the *Gens de la mer*, you shall see how active I will be, and how I will crow over the *Gens du nord*," the name by which the Athabasca voyagers are designated No fresh water was found on Garry Island until Augustus discovered a small lake, the streams that poured down from the cliffs being as salt as the sea. The temperature of the sea water was 51°; the fresh water we had left at five miles from the island 55°; and that of the air 52°.

Garry Island is about five miles long, by two broad, and seems to be a mass of frozen mud, which, in the parts exposed to the air and sun, has a black earthy appearance. It is terminated to the north-west by a steep cliff, through which protrude, in a highly-inclined position, several layers of wood-coal, similar to that found in the Mackenzie. There was likewise observed a bituminous liquid trickling down in many parts, but particularly near the south-west point of the cliff where the bank had been broken away, and a hollow cavity was formed. The ravines and gullies were still filled with ice, though none was seen on the level ground. There were no stones above the sea level; those on the beach consisted of granite, greenstone, quartz, and lydian-stone, of a small size and completely rounded. The vegetable productions were grasses, a few mosses, and some shrubs, the latter in flower. Four foxes were the only land animals we saw;

and a small hawk, some gulls, dotterels, and phaleropes, composed the list of birds. A large medusa was found on the beach.

The sky was cloudless on the morning of the 17th, which enabled us to ascertain the position of our encampment to be in latitude 69° 29′ N., longitude 135° 41′ W., and the variation of the magnetic needle to be 51° 42′ E. We likewise found that it was high water that day at one P.M., with a rise and fall of eight inches, but the direction of the flood could not be ascertained. I wrote for Captain Parry an account of our progress, with such information as he might require, in case he wished to communicate either with the Company's Post at Fort Good Hope, or our party, and deposited my letter, with many others that I had in charge for himself and the officers of the ships, under a pole erected for the purpose, on which we left a blue and red flag flying, to attract his attention. Another statement of our proceedings was encased in a water-proof box, and committed to the sea, a mile to the northward of the island. The wind blew strong off the land at the time, and there was a gale from the northwest the next day, so that there is every chance of the letter having made good way to the eastward.

Having completed the observations, we embarked at two P.M., and pulled along the western shore of the island three miles to the sandy spit at its south-west end, on which there was a vast quantity of drift-wood piled by the action of the waves. From this point we launched forth to cross towards the Mackenzie under double-reefed sails, as the wind was blowing strong, and the waves high in the offing; but finding the boat very stiff and buoyant, the sail was increased, and reaching the eastern point of Ellice Island by seven P.M., we encamped at the foot of the outermost of the three hummocks mentioned on the 15th of August. As we passed along the shore of the island, we disturbed some moose and rein deer, and several geese, cranes, and swans, that were quietly feeding near the water. At this period of the year, therefore, there would be no lack of food, in this country, for the skilful hunter. In the course of the evening I found that a piece of the wood-coal from Garry's Island, which I had placed in my pocket, had ignited spontaneously, and scorched the metal powder-horn by its side.

Our enterprising precursor, Sir Alexander Mackenzie, has been blamed for asserting that he had reached the sea, without having ascertained that the water was salt. He, in fact, clearly states that he never did reach the salt water. The danger to which his canoe was exposed in venturing two or three miles beyond Whale Island, (which lies to the eastward of our route,) at a time when the sea was covered with ice to the north, is a sufficient reason for his turning back; and we can abundantly testify that those frail vessels are totally unfitted to contend against such winds and seas as we experienced in advancing beyond the volume of fresh water poured out by the Mackenzie. It is probable, therefore, that even had the sea been free from ice at the time of his visit, he could not have gone far enough to prove its saltness, though the boundless horizon, the occurrence of a tide, and the sight of porpoises and whales, naturally induced him to say that he had arrived at the ocean. The survey of the Mackenzie made on this Expedition, differs very little in its outline from that of its discoverer, whose general correctness we had often occasion to admire. We had, indeed, to alter the latitude and longitude of some of its points, which he most probably laid down from magnetic bearings only; and it is proper to remark, that in comparing our magnetic bearings with his, throughout the whole course of the river, they were found to be about fifteen degrees more easterly; which may, therefore, be considered as the amount of increase in variation since 1789. In justice to the memory of Mackenzie, I hope the custom of calling this the Great River, which is in general use among the traders and voyagers, will be discontinued, and that the name of its eminent discoverer may be universally adopted.

The excursions to Garry Island having made us acquainted with the state of the sea to the northward, and having shewn that, the bank at the mouth of the river being passed, there was no visible impediment to a boat's proceeding eastward, I was desirous of making further examination in aid of the future operations of the Expedition, by going over to the western shore, and of reaching, if possible, the foot of the Rocky Mountains. With these intentions we embarked at nine A.M., but before we could get half way to the nearest part of Pitt Island, a gale of wind came on from N.W., followed by violent squalls, which, from the threatening appearance of the clouds, and the rapid descent of the thermometer from 68° to 51°, seemed likely to be of some continuance. The design was, therefore, abandoned, and the boat's head directed towards the entrance of the river. It proved, however, no easy task to get into the proper channel; and to effect this object the officers and crew had to drag the boat half a mile over a bar, while the waves were beating into it with such force as to make us apprehensive of its being swamped. As soon as we were in deep water, all the sail was set that the boat could bear, and at two P.M. we arrived at the narrow part. Here, likewise, the waves were high and breaking, and for the purpose of avoiding these and the strength of the current, we kept as close to the shore as possible, going through the water at seven miles an hour, and about four over the current. The wild fowl, warned by the sudden change of the weather, took advantage of this fair wind, and hastened away in large flights to the southward. At ten P.M., the boat having twice grounded, from our not being able to see our way clearly, we halted to sup, and laid down to sleep before a good fire. Temperature at 45°.

When day-light permitted us to distinguish the channels, we embarked again, and scudded under the foresail before the gale, which this day blew with increased violence. We halted to breakfast near some winter habitations of the Esquimaux, which we supposed, from the freshness of the wood-shavings, and the implements of fishing that were scattered about them, had been abandoned only in the preceding spring; and as it was probable they would revisit this spot, we fixed to the pole of a tent a present of a kettle, knife, hatchet, file, ice-chisel, some beads, and pieces of red and blue cloth. These huts were constructed of drift wood, in a similar manner to those which will be described in a subsequent part of the narrative. A second present was deposited at some other huts, and a third at those below the rapids. We imagined that some, if not all, of these would be found by the Esquimaux, and would make them acquainted with our visit. By noon we had advanced as far as the rapid, which we ascended under sail; and at a few miles above this point, owing to the fogginess of the atmosphere, we took a more western channel than that by which we had descended. This proved circuitous, though it ultimately brought us to the former route. It was quite dark before we could find a secure place for the boat, and a sheltered spot for the tent. The gale continued without abatement, the weather was raw and cold, and it was with difficulty we collected some sticks to kindle a fire. Temperature 40°.

On the 20th the wind was moderate. We resumed our journey at four A.M.; passed our sleeping-place of the 12th by noon, and at sunset encamped at the narrow part of the river where the numerous channels commence. Large flights of geese and swans were observed passing to the southward all this day.

G

The musquitoes again made their appearance, though the temperature was at 45°: scarcely any of them had been seen on the descent to the sea.

Sunday, 21st.

Temperature at daylight, on the 21st, 37°. We commenced our labour under oars, but a strong gale from the southward soon rendered this mode of ascending the river ineffectual. The men were, therefore, divided into two parties, who towed the boat by line, relieving each other at intervals of an hour and a half. At fifty minutes past one P.M. we were abreast of the Red River, and there met a large party of the lower Loucheux Indians, who had assembled to wait our arrival. They welcomed our return with every demonstration of joy, more particularly that of Augustus and Baptiste, and at first cheerfully assisted the men in towing, but, like Indians in general, they soon became tired of this labour, and rather impeded than forwarded our progress. So we distributed to each a present; made known as well as we could by signs, that at our next visit we would purchase whatever fish or meat they might collect, and took our leave of them. Owing to the detention these men and another party occasioned, we were caught by a heavy gale from N.W. before we could reach our encampment at the head of the Narrows, and had to pitch the tent in pelting rain. Temperature 43°.

Monday, 22nd.

On the 22nd, we started at four in a thick wet fog, which gave place to snow and sleet, and sailed the whole day before a strong N.W. wind, much to the annoyance of several Indians who tried to keep pace with the boat, by running along the shore: each of them had a present of tobacco thrown to him. We encamped near the bottom of Cannon-shot Reach; the weather was extremely cold, and, during the night, ice was formed in the kettle.

Tuesday, 23rd.

On the next day the wind came contrary from S.E., which obliged us to have recourse to the tow-line. The frequent recurrence of sand-banks, to avoid which we had either to pull round or cross the river, made this day's operations very tedious. In turning round one of the points, we came suddenly upon a party of Indians, who had not seen us on our way down. Our appearance, therefore, created great alarm; the women and children were instantly despatched to the woods, and the men came down to the beach with their guns and arrows prepared and knives drawn; but the explanation that Baptiste gave, soon allayed their fears. They were, indeed, objects of pity; all their property had been destroyed to testify their grief at the death of some of their relations, and the bodies of several were still sore from the deep gashes they had inflicted on themselves in their demonstrations of sorrow. We distributed such useful articles among them as we had remaining, but the supply was not at all equal to their necessities. Several of them attempted to follow us in their canoes by poling, which they dexterously perform by pushing at the same time with a pole or paddle in each hand; the boat, however, was towed faster than they could ascend the stream, and they were soon far behind. We arrived at six P.M. at the Trading River, and there met another party of the Loucheux, among whom was the woman whose tears had excited our sympathy on the 11th, now in high glee, and one of the most importunate for beads. The boy was likewise there to whom the tea had been given as a remedy for his fever, completely recovered, which was, no doubt, ascribed to the efficacy of the medicine. Not choosing to encamp near these people, we crossed the river, and towed four hours longer, when we reached Fort Good Hope. Mr. Dease, and all at his fort, were overjoyed on seeing us again, because the Indians had begun to surmise, and in fact had brought a report that we had all been massacred by the Esquimaux; and had we been detained another week, this statement would have gained entire credence, and, in all probability, spread throughout the country.

The Indian whose fish we carried on our way down, happened to be at the fort, and he cheerfully communicated, through the interpreter, a female, all the information that he or his tribe possessed respecting the mouth of the river, the sea-coast, and the Esquimaux; all topics highly interesting to us, but we subsequently found that his knowledge of these matters was very imperfect. We made known to him our wish that the Esquimaux should be informed of our arrival as soon as possible, and signified that a very substantial present would be given to any person that would carry the intelligence to them in the course of the following winter. Mr. Dease pressed this point strongly on his consideration. This gentleman, indeed, was anxious to promote our desires in every respect, and promised that his utmost exertions should be used to procure a good supply of provision for our next summer's voyage, though he represented the hunters in this vicinity as unskilful and inactive, and begged of me not to rely too much on his collection. We left in his charge five bags of pemmican, and the superfluous stores, to lighten the boats. We quitted the fort in the afternoon with a contrary wind, and towed twenty miles up the stream before we encamped, though the beach was composed of sharp stones, which rendered walking very unpleasant.

The wind being contrary during the four following days, we could only ascend the river by using the tracking line. Our crew cheerfully performed this tedious service, though three of them had been much reduced by dysentery, brought on by previous fatigue, exposure to wet, and by their having lived for some time on dried provision. These men, however, had gradually been gaining strength since the fresh meat was procured on Ellice Island.

On the 25th we came to the aspin, poplar, and larch, in latitude 67° 10′ N., and were not a little surprised to observe the change in their foliage within the last fortnight. Their leaves had assumed the autumnal tint, and were now fast falling. The wild fowl were hastening in large flocks to the south, and every appearance warned us that the fine season drew near its close.

In the passage through the rampart defile, several families of the Hare Indians were observed encamped on the heights, for the purpose of gathering berries, which were at this time ripe, and in the best flavour. At the first sight of the boat the women and children scampered down wherever descent was practicable, to get at their canoes, that they might cross over to us, but we travelled so fast that only a few could overtake the boat. The Indians who reside near this river, from their want of skill in hunting, principally subsist, from spring to autumn, on the produce of their fishing nets, and on wild berries. At the influx of small streams, or wherever there is any eddy, a net is set. In shallow water it is suspended upon sticks planted in a semicircle, so as to enclose the mouth of the river, or the sweep of the eddy; but where the water is deep, and the shore bold or rocky, two stout poles are firmly secured at a short distance from the water's edge, the breadth of the net apart, to the ends of which pliable rods are fastened, of a length sufficient to hang over the water, and to these the net is attached. In the winter these Indians snare hares, which are very abundant in this quarter.

On the 29th we arrived at the upper rapids, which were scarcely discernible at the time of our descent; but from the falling of the water since that time, there was a dry sand-bank of considerable extent in the centre, and the waters on each side of it were broken and covered with foam. Augustus being tired of tracking, had wandered from us to the extremity of this bank, from whence he could not be extricated without great hazard, unless by making him return to the bottom of the rapid. As

1825.
August.
this, however, would have compelled the poor fellow to pass the night upon the sand-bank, Mr. Kendall undertook to bring him off, by running with the current to the point at the commencement of the rapids, which he effected in a masterly manner, although the boat struck twice, and was in considerable danger from the violence of the eddies.

We found, at the place of our encampment, a solitary old woman, sitting by a small fire, who seemed somewhat alarmed at her visitors, until she was joined, after dark, by her husband and son. As soon as the man understood from our signs that we were desirous of having some fish for supper, he instantly embarked to examine his nets; but as they proved to be empty, the woman generously dragged a pike out of a bundle on which she was sitting, and presented it to us, though it was evidently reserved for their own meal. In return we furnished them with a more substantial supper, and made them some useful presents. The weather was extremely sultry throughout this day; at two P.M. the thermometer stood in the shade at 66°, and at 76° when exposed to the sun. The refraction of the atmosphere, which we had often remarked to be unusually great since we had entered the Mackenzie, was this day particularly powerful. The mountains were distorted into the most extraordinary shapes, and the banks of the river, which we knew to be only from thirty to sixty feet high, appeared to have such an elevation, that it would have been impossible for us to recognise the land. The air became cooler in the evening, and the atmosphere less refractive. Soon after sunset the objects appeared in their proper form, and we enjoyed the prospect of the delightful mountain scenery that distinguishes this rapid.

Tuesday, 30th.
Favoured by a N.W. gale, we made great progress on the 30th. The temperature of the air varied in the course of the day from 62° to 41°. The brulôts and sand-flies were very

1825.
August.

teasing wherever we landed; but these, unlike the musquitoes, disappear with the sun.

The upper parts of the Rocky Mountains on the western side of the river were, at this time, covered with snow, but not those on the eastern side, which are, probably, less elevated than the former. We had no opportunity of ascertaining their height, though we conjectured that the loftiest did not exceed two thousand feet, as it was free from snow in the early part of August.

September,
1st.

At sunset this evening we quitted the muddy waters of the Mackenzie, and entered the clear stream that flows from the Great Bear Lake; but owing to the shallowness of the water near its mouth, and the beach being a mere collection of stones, we had to grope our way long after dark in search of a place for an encampment, stumbling and falling at every step. At length we espied a light about a mile further up the river on the opposite shore; we, therefore, crossed over, at the expense of some heavy blows to the boat, and tracked along the base of a steep bank, until we reached the fire. There we found a Canadian and two Indian boys who had been sent from Bear Lake three days before in a canoe, to procure some white mud from the banks of the Mackenzie to decorate our houses. This man was the bearer of a letter from Lieutenant Back to me, which detailed the proceedings at the Fort.

Friday,
2nd.

We embarked at daylight, having the canoe in company. The weather was cold and raw throughout the day; the temperature from 34° to 45°; but the party were kept in constant exercise, either in tracking or walking; the steersman and bowman only being required in the boat. Except where the river was bounded by steep cliffs, the path was pretty good. Its general breadth varied from three hundred to five hundred yards, and its banks were tolerably well wooded, but the trees were small.

This morning the ground was firmly frozen, and the thermo-meter stood at 28°, when we commenced our operations. Early in the afternoon we arrived at the lower part of the mountain that is delineated in the annexed engraving, and which we had kept in view this day, and the greater part of the preced-ing. As we had now to ascend a succession of rapids for fifteen miles, and two of our crew were lame, I directed the canoe to be laid up on the shore, and took the Canadian and the boys to assist at the tow-line. We had not advanced more than two miles before we met with an accident that was likely to have been attended with serious consequences: in the act of hauling round a projecting point, and in the strength of the current, the tow-line broke, and the boat was driven with great force against a large stone at some distance from the shore, having deep water on every side. There it lay with the broadside exposed to the whole pressure of the current, beating violently against the stone; and from this situation it could not have been extricated, had not Gustavus Aird, the strongest man of the party, ventured to wade into the river, at the imminent risk of being swept off his feet, until he could catch the rope that was thrown to him from the boat. As soon as it was dragged to the shore, we found that part of the keel was gone, and the remainder much twisted, and all the fastenings of the lowest plank were loosened. The carpenter set to work to repair this mischief in the best manner he could with the materials he had, and before night the boat was again launched. The leaks, how-ever, could not be quite stopped, and in our further progress one of the men was constantly employed baling out the water.

The next day's operations were tedious and hazardous as long as the rapids continued. The men had to walk with the tow-line along a narrow ledge that jutted out from the base of a

1825.
September.

steep rocky cliff, which was very slippery from the rain that had fallen in the night: a false step might have proved fatal; and we rejoiced when, having passed the rapids, we found earthy banks and a better path. The services of Augustus and the Indian lads being no longer required, I despatched them to the Fort, to apprize the party there of our approach.

We had a severe frost this night: at daylight in the morning the thermometer was down to 20°, and a raw fog contributed to make the weather very cold and comfortless. The sun shone forth about eleven, and soon dispersed the fog, and then the temperature gradually rose to 54°.

At four P.M. we arrived at the foot of the upper rapid, and in two hours afterwards entered the Great Bear Lake, and reached the house at seven. Dr. Richardson having returned from his voyage to the northern part of the lake, the members of the Expedition were now, for the first time, all assembled. We heartily congratulated each other on this circumstance, and also on the prospect of being snugly settled in our winter-quarters before the severe weather. Dr. Richardson had surveyed the Bear Lake to the influx of Dease's River, near its N.E. termination, at which point it is nearest to the Coppermine River. He fixed upon the first rapid in Dease River as the best point to which the eastern detachment of the Expedition could direct its steps, on its return from the mouth of the Coppermine River the following season. The rapid was, by observation, in latitude 66° 53′ N., and longitude 118° 35′ W., and the variation of the magnetic needle there, was 47° 29′ E.

Monday,
5th.

H

THE FOLLOWING TABLE CONTAINS THE DISTANCES TRAVELLED
BY THE EXPEDITION DURING THE SUMMER OF 1825.

PRINCIPAL PLACES.	STATUTE MILES.
From New York to Penetanguishene, by the route we travelled. , . . .	760
Lake Huron	250
Lake Superior	406
From Fort William to Cumberland House	1018
Cumberland House to Fort Chipewyan	840
Chipewyan to Fort Resolution, Slave Lake	240
Fort Resolution to the commencement of the Mackenzie	135
Head of the Mackenzie to Fort Simpson	103
Fort Simpson to Bear Lake River	271
Bear Lake River to, and the return from, Garry Island.	1206
Length of the Bear Lake River to the Fort.	91
Dr. Richardson's excursion to the north-east termination of Bear Lake.	483
Distance travelled	5803
Number of Miles surveyed	2593

CHAPTER II.

TRANSACTIONS AT FORT FRANKLIN, 1825-26.

Mr. Dease having passed the winter of 1824-25 at the Big Island of Mackenzie, arrived here with fifteen Canadian voyagers, Beaulieu, the interpreter, and four Chipewyan hunters, on the twenty-seventh of July, 1825; which, on account of the drifting of the ice, was as soon as he could, with safety, ascend the Bear Lake River. Several of the Dog-Rib Indians were on the spot, which enabled him to take immediate steps towards procuring a supply of dried meat for our winter use, as well as of fresh meat for present consumption. It having been ascertained that the Rein-deer are most abundant in the north-east quarter of the lake, during the months of August and September, a select party of Indians was despatched to hunt thereabout, under the direction of the interpreter, who took a large canoe for the purpose of bringing home the produce of their hunt. Other men were sent to inform the Hare Indians of our wish to purchase any meat they might bring to the establishment. Our principal subsistence, however, was to be derived from the water, and Mr. Dease was determined in the selection of the spot on which our residence was to be erected, by its proximity to that part of the lake where the fish had usually been most abundant. The place decided upon was the site of an old fort belonging to the North-West Company, which had been abandoned many years;

1825.
September.

H 2

our buildings being required of a much larger size, we derived very little benefit from its materials. The wood in the immediate vicinity having been all cut down for fuel by the former residents, the party was obliged to convey the requisite timber in rafts from a considerable distance, which, of course, occasioned trouble and delay. We found, however, on our arrival, all the buildings in a habitable state, but wanting many internal arrangements to fit them for a comfortable winter residence. They were disposed so as to form three sides of a square, the officers' house being in the centre, those for the men on the right, with a house for the interpreter's family, and the store on the left. A blacksmith's shop and meat store were added, and the whole was inclosed by the stockading of the original fort, which we found highly serviceable in skreening us from the snow-drift and wintry blasts. The officers' dwelling measured forty-four feet by twenty-four, and contained a hall and four apartments, beside a kitchen. That of the men was thirty-six feet by twenty-three, and was divided into three rooms. These buildings were placed on a dry sandy bank, about eighty yards from the lake, and twenty-five feet above it; at the distance of half a mile in our rear, the ground rose to the height of one hundred and fifty feet, and continued in an even ridge, on which, though the timber had been felled, we found plenty of small trees for fuel. This ridge bounded our view to the north; and to the west, though confined to less than two miles, the prospect was pretty, from its embracing a small lake, and the mouth of a narrow stream that flowed in at its head. Our southern view commanded the south-west arm of Bear Lake, which is here four miles wide, and not deeper than from three to five fathoms, except in the channel of the river, which conveys its waters to the Mackenzie. We had also, in front, the Clark-hill, a mountain about thirty-six miles distant, which was always visible in clear weather. When the refraction was great, we saw the tops of some other hills, belonging to the range that extends from Clark-hill to the rapid in Bear Lake River.

Immediately under the sandy soil on which the house stood, there is a bed of tenacious bluish clay, of unknown thickness, which, even in the months of August and September, was firmly frozen at the depth of twenty-one inches from the surface. No rocks were exposed in any part, and wherever the surface had been torn up, a clayey soil appeared. Many boulder stones of granite, lime-stone, sand-stone, and trap rocks, were scattered about the lake, not far from the shore.

The trees at some distance from our fort consisted of black and white spruce, and larch, generally small, though a few of the better grown measured from four to five feet in girth, and were from fifty to fifty-five feet high. Dr. Richardson ascertained, by counting the annual rings, that some of them, in a sound state, were upwards of one hundred and thirty years old; while others, which were not much greater in size, had two hundred and fifty rings, but these were decayed at the heart.

The officers had done me the honour, previous to my arrival, of giving the name of Franklin to the fort, which I felt a grateful pleasure in retaining at their desire, though I had intended naming it Fort Reliance. The number of persons belonging to the establishment amounted to fifty: consisting of five officers, including Mr. Dease; nineteen British seamen, marines, and voyagers; nine Canadians; two Esquimaux; Beaulieu, and four Chipewyan hunters; three women, six children, and one Indian lad; besides a few infirm Indians, who required temporary support. This party was far too large to gain subsistence by fishing at one station only; two houses were, therefore, constructed at four and seven miles distance from the fort, to which parties were sent, provided with the necessary fishing implements; and not more than thirty persons were left to reside at the principal establishment. From fifteen to twenty nets were kept in use, under the superintendence of Pascal Coté, an experienced fisherman, who had two assistants. These were placed opposite the house, and towards the end of summer, and in autumn, they yielded daily from three to eight hundred fish, of the kind called " the Herring Salmon of Bear Lake," and occasionally some trout, tittameg, and carp. Four Dog-Rib Indians, who were engaged to hunt the Rein-deer in the neighbourhood of the fort, from want of skill, contributed very little fresh meat to our store. Augustus and Ooligbuck employed themselves in the same service, but from not being accustomed to hunt in a woody country, they were not more successful.

The consideration of next importance to furnishing the party with food, was to provide regular occupation for the men, who had not the resources to employ their time which the officers possessed. Accordingly, some were appointed to attend exclusively to the fishing nets, others to bring home the meat whenever the hunters killed any deer; some were stationed to fell wood for fuel, others to convey it to the house, and a third set to split it for use. Two of the most expert travellers on snow-shoes were kept in nearly constant employment conveying letters to and from the posts in the Mackenzie and Slave Lake. As the days shortened, it was necessary to find employment during the long evenings, for those resident at the house, and a school was, therefore, established, on three nights of the week, from seven o'clock to nine, for their instruction in reading, writing, and arithmetic; and it was attended by most of the British party. They were divided in equal portions amongst the officers, whose labour was amply repaid by the advancement their pupils made: some of those who began with the alphabet, learned to read and write with tolerable correctness. Sunday was a day of rest; and, with the exception of two or three of the Canadians, the whole party uniformly attended Divine service, morning and evening. If, on the other evenings for which no particular occupation was appointed, the men felt the time tedious, or if they expressed a wish to vary their employments, the hall was at their service, to play any game they might choose; and on these occasions they were invariably joined by the officers. By thus participating in their amusements, the men became more attached to us, at the same time that we contributed to their health and cheerfulness. The hearts and feelings of the whole party were united in one common desire to make the time pass as agreeably as possible to each other, until the return of spring should enable them to resume the great object of the Expedition.

The officers found employment in making and registering the thermometrical, magnetical, and atmospherical observations, which were hourly noted from eight A.M. to midnight; and, in addition to the duties which they had in common, each had a peculiar department allotted to him.

Lieutenant Back had the superintendence of the men; and the accurate drawings which he finished during the winter, from sketches taken on the voyage, afford ample proof of his diligence and skill. Dr. Richardson, beside the duties of medical officer, which, from the numerous applications made by the natives, were not inconsiderable, devoted his attention to natural history, as well as to a series of observations on the force of the sun's radiation. Mr. Kendall constructed all the charts after the data had been recalculated by myself; he also made several drawings; and he undertook an interesting series of observations on the velocity of sound, which will appear in the Appendix. To Mr. Dease the charge was committed of whatever related to the procuring and issuing of provision, and the entire management of the Canadian voyagers and Indians.

Previous to the officers leaving London, Dr. Fitton, President of the Geological Society, had the kindness to devote much of his time to their instruction in geology; and having furnished them with a portable collection for the purpose of reference on the voyage, Dr. Richardson, when he had leisure, explained these specimens, weekly, to the party, and assisted them in reading on this science, which proved a most agreeable and useful recreation to us all.

Some of the preceding remarks refer to a period of our residence later than that which I am about to enter upon; but I thought it best to insert them here, that the mention of them might not interrupt the narrative of occurrences which I shall now resume, reserving detailed notices of the appearance of the Aurora Borealis, of the deviations of the magnetic needle, and the observations on temperature, for the Appendix.

Thursday,
8th.

On September 8th, two men were sent off to Slave Lake, in a canoe, with a despatch, containing an account of our proceedings, addressed to His Majesty's Secretary of State for the Colonies; and as we expected letters from England, by the way of Hudson's Bay, they were directed to await their arrival at Slave Lake. There was almost constant rain from the 11th to the 14th, which much retarded the work going on out of doors, and particularly the construction of an observatory, which we were desirous of completing as soon as possible, that the magnetical observations might be commenced. We found employment, however, in whitewashing and fitting up the interior of the different houses. The 15th proving fine, we established a meridian line, and ascertained the variation by each of the compasses.

Tuesday,
20th.

Beaulieu returned with his family, the Chipewyan hunters,

and some Dog-Ribs, bringing a supply of dried meat, rein-deer tongues, and fat, sufficient for a month's consumption, which was reserved for use when the fishing should become unproductive. These men reported, that at the time they quitted the northern shores of the lake, the deer were retiring towards this quarter; which intelligence accounted for the Indians having killed four within a day's march from the house. 1825. September.

The chimney of the last of the buildings being completed this morning, the flag-staff erected, and all the men assembled, we commemorated these events by the festivities usual on the opening of a new establishment in this country. The first part of the ceremony was to salute the flag; the men having drawn themselves up in line, and the women and children, and all the Indians resident at the fort, being disposed in groups by their side, a deputation came to solicit the presence of the officers. When we appeared, we found our guns ornamented with blue ribbons, and we were requested to advance and fire at a piece of money which was fastened to the flag-staff. The men then fired two volleys and gave three hearty cheers, after which Wilson the piper struck up a lively tune, and placing himself at the head of his companions, marched with them round to the entrance of the hall, where they drank to His Majesty's health, and to the success of the Expedition. In the evening the hall was opened for a dance, which was attended by the whole party, dressed in their gayest attire. The dancing was kept up with spirit to the music of the violin and bag-pipes, until daylight. Friday, 23rd.

These entertainments over, Beaulieu and the hunters were despatched to the chace, and they soon added two moose-deer to our store. Monday, 25th.

There had been much rain in the course of the preceding week, and the temperature was generally mild, but a fall of snow Tuesday, 25th.

took place on the 27th. Some Dog-Ribs came to the fort on that day with the produce of their autumnal hunt, which was very inconsiderable, but they rendered good service to us by taking away with them several of their relations, who had been subsisting on our bounty for some time. After their departure there only remained one man of the tribe, who, being afflicted with rheumatic fever, was retained under the care of Dr. Richardson. Warm clothing was provided for him, and a comfortable leathern lodge was erected for himself and family.

The month of October commenced with frost and snow, and the party were now furnished with fur caps, leathern mittens and trowsers, and the rest of their warm winter-clothing. This day we completed the erection of the observatory, and adjusted an instrument to the magnetic meridian, for the purpose of observing the variations of the needle.

Much snow fell on the night of the 7th, and on the 11th the small lake was firmly frozen over, and the ground in the same state. All the migratory birds being now gone, except a few ducks, which still lingered in the open water of Bear Lake, we considered this day to be the first of the winter. It was remarkably clear and fine, and we hailed the commencement of this season with a degree of pleasure, from its contrast with the wet unsettled weather which marks the close of summer. A few clouds passing over the sun's disk, produced an instantaneous depression of ten degrees of the mercury in a thermometer exposed to the sun's rays. The atmospherical refraction was remarkably strong at this time. We had repeated opportunities, in the course of the winter, of observing it to be greatest in similar states of the atmosphere.

The boats were now secured for the winter in a sheltered place, and screened as much as possible from the effects of the wind and snow drift, by a strong fence made of boughs and branches.

We were surprised on the 14th by the arrival of two Canadians from Fort Norman, with letters from Governor Simpson, and other gentlemen in the southern districts, containing satisfactory answers to the requisitions for stores that I had made in my passage through the country. We were also pleased to learn that Thomas Matthews, the carpenter, whom we had left at Cumberland House, on account of his leg being broken, had reached Fort Norman, in the Company's canoe; and I felt much indebted to Mr. James Keith, and Mr. Smith, Chief Factors, for the care and tenderness with which they had conveyed him through the country.

The season at which the ice begins to form, is the most favourable for fishing in the lakes of this country, and we then procured from four to five hundred daily. Those not required for immediate consumption, were hung on a stage to freeze, in which state they keep until the following spring. But we could not derive the full advantage from the season, because the drift ice, making it unsafe to keep the nets set in Bear Lake, they were taken up on the 18th. Near a month elapsed before they could be set with safety under the ice; our first attempts resulting in the loss of three nets. We procured, however, a few fish from the small lake, during this interval, and the rest of our food was supplied from the store of dried meat.

We were visited on the 20th by a storm of snow, which continued, without intermission, for thirty-six hours. Although it put an end to the skating, and the games on the ice, which had been our evenings' amusement for the preceding week, yet the change made every one glad, because the snow was now deep enough for winter travelling. We had learned, some days before, that the hunters had stored fifteen rein-deer in the woods, and on the 22nd four men were despatched with sledges to bring them to the fort.

I 2

The first throw off of the dog-sledges for the season never fails to attract general attention; accordingly the whole party was collected to witness it on this occasion. They set off at full speed, and were soon out of sight. From this time dog-sledges were used to drag the fuel, which had been hitherto done by the men. We sent a party to cut down timber, and saw it into planks, fit for the construction of another boat.

Wednesday, 26th.
On the 26th the thermometer first fell below zero, but the month closed with a very calm, mild day. Mr. Kendall and I were employed in measuring a geographical mile on the small lake, preparatory to a series of observations on the velocity of sound. The only ferine companions we now had were a few hardy quadrupeds and birds, capable of enduring the winter. The variety of the former was confined to wolves, foxes, martens, hares, mice, and a few rein-deer. Of the feathered tribe, there were the raven and Canadian crow, some snow-birds, wood-peckers, red-caps, crossbeaks, Canada, rock, and willow partridges, and a few hawks and owls.

November.
Having received information that the Hudson's Bay Company intended sending their annual despatch from the Mackenzie River to York Factory, by the close of this month, and the ice

Wednesday, 9th.
on Bear Lake and the Mackenzie River being, on the 9th, sufficiently strong, we forwarded a packet of letters to Fort Norman, and a dog-sledge to convey Thomas Matthews to this place. On the 15th the nets were reset under the ice, and we were relieved from the necessity of putting the party on short allowance. We had the additional pleasure of learning that the hunters had killed ten rein-deer. The men returned from Fort Norman on the 18th, accompanied by Thomas Matthews, whose leg was yet too weak for him to walk more than a short distance.

During the middle, and towards the close of November,

parheliæ were frequent; the most brilliant appeared on the 27th; it continued as long as the sun was above the horizon. The atmosphere was cloudless, and apparently free from haze, except just about the sun, which seemed to gleam through a fog. The surrounding circle was nearly complete, and displayed the prismatic colours vividly; from the centre of the sun's disk a beam of bright light extended upwards several degrees beyond the circle. The inner radius of the circle measured 21° 34′, and the outer 22° 50′. The wind blew fresh all the day from E. N. E., and the temperature was 10°. In the evening the moon was encircled by two distinct halos; temperature 7°.

This morning the principal leader of the Dog-Ribs, and a large party of his tribe, came to the fort. It is usual for Indians, on the first visit to an establishment, to make their approach in line, and with much formality; but on this occasion our visitors shewed an unusual degree of caution. Their distrust had originated in a very trifling occurrence at the close of our house-warming festivities on the 23d of September. Some of the Canadians having asked Mr. Dease if our Highlandmen did not come from the same country with the rest of the English party, were told that they were natives of the mountainous lands, or *Montagnards*. This name unfortunately being used by the voyagers to designate the Dog-Ribs, was considered by the Highlanders to be a term of reproach when applied to themselves, and a scuffle ensued. Harmony was soon restored by the officers sending the most noisy to bed, and next morning the true meaning of the word Montagnard was explained to the Highlandmen, and the party set about their usual occupations with their wonted good feeling towards each other. Not so with an unlucky Dog-Rib, who had been attracted to the scene by hearing the name applied by the voyagers to his countrymen bandied about from one to the other, and thrusting his head into the crowd had received a blow. This at once confirmed all his fears, and he fled to spread a report amongst his countrymen that the white people intended to destroy the Indians. Although his report was not fully believed, yet it produced the feeling of distrust which the Indians manifested on their approach to the house. It was entirely removed by the explanation we gave. These Indians having brought a quantity of furs for the Hudson's Bay Company, as well as dried meat for ourselves, and I having understood from Mr. Dease that it would be an accommodation to them if they were permitted to deposit their furs at this place, instead of carrying them to Fort Norman, I acceded to this suggestion, and directed Mr. Dease to advance from our stores the goods required for the purchase of the furs, which were to be returned when we should visit that fort in the spring.

An old man belonging to the Company's establishment at Fort Norman arrived this day with his wife, to stay some time with us, because the supply of provision had failed at that post. We felt much pleasure in sharing our means with this aged couple, who were much reduced by their late scanty fare.

The close of November was marked by a succession of strong east winds, and a mildness of temperature, rare at this season. On the 30th the thermometer rose from + 18° to 29° above zero, on the occurrence of a gale from the north.

December.
Thursday,
1st.

Friday,
2nd.
The first of December being a cloudless day, we endeavoured to observe the latitude at noon, but failed, owing to the extraordinary atmospherical refraction. At midnight, on the 2nd, there was a shower of hail, so small that we could hardly distinguish it from rain. Dr. Richardson thought he perceived lightning. Temperature + 22°, calm. On the night of the

4th another instance of a sudden increase of temperature from + 7° to + 26° was observed, on a north wind succeeding a calm.

The fishery having gradually declined for some days, our nets were removed nearer to the entrance of Bear Lake River, where the current continued to keep the water open for a considerable space. We then procured a daily supply of fish sufficient for the rations of the household, as well as the dogs, though our number was now increased by the party from the more distant fishery, which had proved unproductive. The allowance was seven of the herring salmon to a man per day, and two to each dog.

The shortness of the days now precluding the Indians from hunting, many came, according to their custom, to spear fish at the head of Bear Lake River, and their numbers gradually increased. They were not, however, successful, nor diligent, preferring to beg what they could from us, and sending their women and children to subsist on the offal of the fish used at the fort. To encourage them to greater exertion, I provided them with nets, and other fishing materials, but their indolence led them to make a very ungrateful return; for on several occasions they emptied our nets in the night, and thus not only robbed us of what they took away, but, by deranging the nets, deprived us of the whole of that day's supply. We never could ascertain the perpetrators of these thefts. The blame was invariably thrown on some aged and infirm men, who denied it. Notwithstanding the straits to which they became reduced, they could not be persuaded to go off to a more productive fishery, until we were compelled to withhold all supplies, from fear of starving our own party. These Indians shewed more indolence, and less regard for truth and honesty, than any other tribes with which we had dealings. Their sufferings are often extreme, and some of them perish every year from famine; although, from the abundance of fish in this country, but slight exertion would be required to lay up, at the proper seasons, a stock for the whole year.

The difficulty of procuring nourishment frequently induces the women of this tribe to destroy their female children. Two pregnant women of the party then at the fort, made known their intention of acting on this inhuman custom, though Mr. Dease threatened them with our heaviest displeasure if they put it into execution: we learned that, after they left us, one actually did destroy her child; the infant of the other woman proved to be a boy. Infanticide is mentioned by Hearne as a common crime amongst the northern Indians, but this was the first instance that came under our notice, and I understand it is now very rare amongst the Chepewyan tribes;—an improvement in their moral character which may be fairly attributed to the influence of the traders resident among them.

On the 18th a party of sixteen Hare Indians, two Copper Indians, and a Loucheux, arrived with sledges of dried rein-deer meat and furs. While the house was in confusion from the unpacking of their lading, a melancholy scene took place, which excited the warmest sympathy. The wife of one of our Dog-Rib hunters brought her only child, a female, for medical advice. As she entered the room it was evident that the hand of death was upon it. In the absence of Dr. Richardson, who happened to be out, all the remedies were applied that were judged likely to be of service; and as soon as he returned, there being yet a faint pulsation, other means were tried, but in vain. So gentle was its last sigh, that the mother was not at first aware of its death, and continued to press the child against her bosom. As soon, however, as she perceived that life had fled, she cast herself on the floor in agony, heightened by the consciousness of having delayed to seek relief till too late, and by apprehension of the anger of her husband, who was doatingly attached to the child. The Indians evinced their participation in her affliction by silence, and a strong expression of pity in their countenances. At the dawn of day the poor creature, though almost exhausted by her ceaseless lamentation, carried the body across the lake for interment.

The 20th being a very stormy day, we were surprised at the arrival of two voyagers from Fort Good Hope, bearers of letters from Mr. C. Dease, conveying the gratifying intelligence that the Loucheux had seen the Esquimaux since the autumn, and that the latter had found the presents which had been left at their huts, and would be delighted to welcome the return of the white people to the Esquimaux lands next spring.

Our constant occupations had made the time pass so swiftly, that the shortest day came almost unexpectedly upon us. The sun rose this morning, (the 22d,) at $10^h 24^m$, thirteen minutes earlier than its appearance was expected from calculation, owing to the great refraction. Mr. Kendall and I measured its meridional altitude from the lake with two instruments, the one bringing its upper limb to the top of the land four miles distant, the elevation of which had been ascertained to be eight minutes, and the other to its base, the depression of which was two minutes. The mean of both these observations, corrected for refraction by the tables in the Nautical Almanack, gave a result of 65° 11′ 56″ N., which latitude exactly corresponds with the best observations made in the preceding autumn. At $8^h 30^m$ P.M. a halo was observed, whose radius measured 28° 40′ from the moon; and at an equal altitude with the latter body there were two paraselenæ, which, as well as the moon, were intersected by a luminous circle, having the

K

zenith for its centre, and a diameter of 94° 15'. The length of our shortest day did not exceed five hours, but the long nights were enlivened by most brilliant moon-light, and we had frequent and very fine appearances of the Aurora Borealis. The latter phenomenon made some of its grandest displays on the 26th of October, the 2nd of November, and the 7th of December. On all these occasions the disturbed motions of the magnetic needle were very remarkable, and a most careful series of observations convinced the party that they had a close connexion with the direction of the beams of light of which the aurora was composed. My observations also led me to conclude that the deviations of the needle were, in a certain degree, connected with changes in the weather; for, previous to a gale or a snow-storm, the deviations were always considerable; but during the continuance of the gale, the needle almost invariably remained stationary.

Preparations were made for the celebration of Christmas. The house was replastered with mud, all the rooms white-washed and repainted, and Matthews displayed his taste by ornamenting a chandelier with cut paper, and trinkets. On the evening of the 24th the Indian hunters' women and children were invited to share in a game of snap-dragon, to them an entire novelty. It would be as difficult to describe the delight which the sport afforded them after they recovered their first surprise, as to convey the full effect of the scene. When the candles were extinguished, the blue flame of the burning spirits shone on the rude features of our native companions, in whose countenances were pourtrayed the eager desire of possessing the fruit, and the fear of the penalty. Christmas Day falling on a Sunday, the party were regaled with the best fare our stores could supply; and on the following evening a dance was given, at which were present sixty persons, including the Indians, who sat as spectators of the merry scene. Seldom, perhaps, in such a confined space as our hall, or in the same number of persons, was there greater variety of character, or greater confusion of tongues. The party consisted of Englishmen, Highlanders, (who mostly conversed with each other in Gaelic,) Canadians, Esquimaux, Chipewyans, Dog-Ribs, Hare Indians, Cree women and children, mingled together in perfect harmony. The amusements were varied by English, Gaelic, and French songs. After these holidays were over, the Dog-Ribs at length yielded to the repeated solicitations of Mr. Dease, and removed in a body to a distant part of the lake, where they now confessed the fishery was more abundant. As the hunters were drawing rations from our store, he despatched them in quest of deer, furnishing them also with nets. After which there remained at the establishment, only one infirm Indian, and his wife.

January 1st, 1826. This morning the men called in the hall to offer the congratulations of the season to the officers, and we afterwards assembled to read divine service. On the evening of the 2nd, similar festivities were held to those at Christmas, to welcome the new year. The temperature was at —49° on the 1st, which was its lowest state during this winter. This severe weather was of short continuance, for on the 3rd there was a storm of snow, and the thermometer rose to —9°.

Accompanied by Mr. Dease, and Fuller, the carpenter, I walked several miles in search of birch-trees fit for the keel and timbers of the new boat. We found some that would answer for the latter purpose, but none for the keel ; we, therefore, substituted pine. The general depth of snow in the woody and sheltered parts was two feet. On the 16th, by the return of the two men who had been sent to Slave Lake, we had the happiness of receiving a packet of letters, which left England in the preceding June. Beside the more interesting private

K 2

communications, our friends had been kind enough to forward piles of newspapers, and several periodical publications. The 'Quarterly Review,' the 'Edinburgh Philosophical Journal,' and a series of the 'Literary Gazette,' and the 'Mechanics Magazine,' were spread upon the table, and afforded us the most agreeable amusement, as well as never-failing topics for conversation. Could any of our friends have dropped in upon us, in the evening, they would have found us discussing the events of the by-gone year, with all the earnestness and interest which we could have shewn had they been the occurrences of the day, and depended upon our decision. This valuable packet had nigh been lost on its way through the interior, owing to the treachery of an Indian. The fellow had undertaken to guide the Canadian servants of the Hudson-Bay Company, who had it in charge, from York Factory to Cumberland House; but supposing, from its being unusual to forward packets at that season, that it must contain something of value, he seized an opportunity, when the two men had gone a little way from the river side, to steal the canoe, with its contents, and cross the river. There were no means of pursuit, and the poor men, destitute of food, without a gun, or even the means of making a fire, were obliged to march to the nearest establishment, through a very rugged and thickly-wooded country. They reached it after many days travelling, and much suffering, and as soon as they arrived, Mr. Mackintosh, the chief of the department, immediately sent off different parties in search of the culprit. They did not find him, though they got possession of the packet, which was torn open, and the letters scattered upon the ground. I need hardly mention that I afterwards remunerated the Canadians for their sufferings and good conduct on this occasion.

On the morning of the 22nd we perceived a gray wolf crossing the lake, and Augustus and Ooligbuck went in pursuit.

The speed of the animal, however, so much outstripped theirs, that it coolly halted to snap up an unfortunate fox which happened to cross its path, and bore it off in triumph. The visits of this animal were repeated for three successive days, and it at last became so bold as to steal fish, on two occasions, from a sledge which the dogs were accustomed to draw home from the nets, without a driver. The dogs were not touched, but this was accounted for when the wolf was killed, and found to be a female, as Mr. Dease informed us that at this season of the year the female wolves never attack the dog.

The month of February was a very anxious period of our winter's residence. The produce of the nets and fishing lines had been gradually diminishing during January, until the supply did not afford more than three or four of the small herrings per man; and none could be furnished to the dogs. The stock of dried meat was expended, and serious apprehensions were entertained of the party's suffering from want of food. The fish too, from being out of season, afforded very little nourishment, and frequent indisposition was the consequence with us all. Three of the stoutest men with whom this diet particularly disagreed, suffered very much from diarrhœa. It became, therefore, necessary to draw upon the stores of provisions which had been set apart for the voyage along the sea-coast, and, on the 6th, we despatched three sledges to Fort Norman, for some pemmican, arrow root, and portable soup: they were likewise to bring any iron that could be procured from that establishment fit for being converted into nails or fastenings for the intended boat. This being the last opportunity of the season for forwarding letters to the southern department, I wrote to Governor Simpson and the Council at York Factory, requesting that supplies of provisions might be stored for the Expedition, on the route to Canada and York Factory, and that the necessary means of conveyance might be provided for its return in 1827. All these arrangements requiring to be made a year in advance, I included the whole party in the estimate of the numbers to be provided for, that there might be no want of provision, if the western part of the Expedition should, from any cause, be obliged to retrace its steps. By the same conveyance I sent an account of our proceedings, with maps and drawings, to be forwarded to the Colonial Office.

On the 4th of this month, when all were heartily tired of short allowance, a report was brought of the traces of a moose deer having been seen about twelve miles from the fort. Had the days been longer, and a crust formed upon the snow, the hunters would have found no great difficulty in running down the animal, but our principal hope lay in their getting within shot without " raising it,"—the expression used when a deer is scared. Beaulieu being the most expert moose-hunter, went out on this occasion, accompanied by two others, Landré a Chipewyan lad, and a Dog-Rib hunter. When they arrived on the deer's track, they found that it had been raised, probably, by the Indians who first discovered it ; but anxious to procure meat for the fort, they commenced the pursuit. From their knowledge of the habits of the animal, and of the winding course it takes, they were enabled to shorten the distance ; but after running four successive days without coming in sight, Beaulieu had the misfortune to fall over the stump of a tree, and sprain his ankle ; the other two hunters being previously tired out. When this accident happened, they knew they were near the deer, and that it would soon give in, because its footsteps were stained with blood. Beaulieu, however, on account of his lameness, returned to the house, and his companions came with him. During the chase they bivouacked on the snow, and subsisted on a few ptarmigan which they killed. Landré, after a night's rest, again set out, and was successful after two more day's running; not, however, without having nearly lost his life, for the moose, on receiving a shot, made a rush at him, striking furiously with its fore feet. He had but just time to shelter himself behind a tree, upon which the animal spent its efforts, until his gun was again ready.

Landré's arrival with the joyful intelligence of his success, was hailed as the commencement of a season of plenty. When the moose meat was brought in, we had not an ounce of provision in store, and it was, therefore, most acceptable; although, from the manner in which it was hunted down, it proved exceedingly tough. In the evening, to increase our satisfaction, an Indian arrived with the information that the fish were plentiful at the station to which the Dog-Ribs had removed, and likewise that the hunters belonging to the fort had killed some rein-deer near their lodges. We immediately equipped four men with nets and lines, and sent them back with the Indian, giving them directions to report whether more persons could gain subsistence there. Their report, a few days afterwards, being favourable, four more men were despatched thither. They sent us some tittameg, weighing from six to eight pounds, which were the more acceptable, because none of that kind had been taken in our nets since the lake had been frozen over. By the time the moose was finished, the men came back from Fort Norman, with three bags of pemmican, which enabled us to continue the daily issue of rations, though the fare was still scanty.

On the 25th, Beaulieu, accompanied by two men, went off in one direction, and the Dog-Rib hunters in another, in search of deer. Both parties were successful. From the former we received a summons, after four days' absence, to send sledges for meat, but not so from the Dog-Ribs, for they, to compensate for their long abstinence, consumed almost all the meat, and gorged themselves to such a degree, that they were unable to move, and became quite ill. From this period we had a sufficient supply of provision, because the fisheries improved, and we received deer from time to time. The men who had been indisposed gained strength, from the increased quantity, and amended quality, of their food; and we had also the gratification of seeing the dogs daily fatten, amidst the general plenty. The conduct of the men during the season of scarcity was beyond all praise; and the following anecdote is worthy of record, as displaying the excellent feeling of a British seaman, and as speaking the sentiments of the whole party. Talking with Robert Spinks as to the difference of his present food from that to which he had been accustomed on board ship, I said I was glad the necessity was over of keeping them on short allowance. " Why, Sir," said he, " we never minded about the short allowance, but were fearful of having to use the pemmican intended for next summer; we only care about the next voyage, and shall all be glad when the spring comes, that we may set off; besides, at the worst time, we could always spare a fish for each of our dogs." During the period of short allowance, the three dogs under the charge of this man were kept in better condition than any of the others.

We now called the men home from the nearest fishery, and set their nets near the Bear Lake River, but the men at the distant station with the Indians were kept there, and occasionally supplied the fort with fine tittameg and trout. The otters did considerable mischief to our nets at this time; six of these animals were seen in one day.

Many parhelia were observed this month. On the 14th, at forty-five minutes after nine A.M., the arched form of the clouds, and the appearance of a collection of rays projected from the sun's disk in the shape of a fan, strongly resembled the coruscations of the aurora. The atmosphere was misty; temperature

1826.
February.

in the shade + 8° 5′; and when the thermometer with a black-ened bulb was exposed to the sun's rays, it rose to + 43°. The magnetic needle, at nine A.M., was perceived to have made a greater deviation to the westward than usual at that hour, and I imagine that the cause of this increase probably arose from the atmosphere being then in a state of electricity, similar to that in which it is when the aurora appears in hazy weather; on which occasions we have observed that its coruscations have the strongest effect in causing aberrations of the needle. A violent gale from the north-west commenced on the 26th, and lasted, without intermission, for thirty-six hours.

March.
Wednesday,
1st.

The early part of this month was marked by a succession of gales from the N.W., with a few intervals of moderate weather, in which the wind came from the east, and was attended by a clearer atmosphere than usually accompanies easterly winds in the colder months. We observed, with pleasure, on the 7th, that the sun had sufficient power to soften the snow in exposed places, and to form icicles from the roofs which had a southern aspect, but the return of strong winds from the W.N.W. brought back severe weather. On the 11th there was a violent gust of wind, which, in its passage over the lake, gathered up the snow in a column, similar to that of a waterspout.

Saturday,
11th.

Dr. Richardson made an excursion for the purpose of examin-ing the rocks to the north of the establishment. He returned after two days, the snow being too deep for him to obtain speci-mens. The description he gave of a view from an eminence nine miles behind the fort, induced Lieutenant Back and me to visit the spot, and we were amply repaid for the walk. The view embraced the mountains on the borders of the Mackenzie to the west, a considerable portion of Bear Lake River, with the mountains near its rapids, Clark's Hill to the south, and the range of elevated land stretching to the east till they were lost in the distance. To the N.E. there appeared several small lakes, and the view was terminated by a portion of Bear Lake.

The Chipewyan hunters who had been absent since Christmas, returned to us with their families, and brought with them a Dog-Rib girl, about twelve years old, who had been deserted by her tribe. When they found her, she was in the last stage of weakness, from famine, sitting by the expiring embers of a fire, and but for their timely appearance, death must soon have ended her sufferings. They fed and clothed her, and waited until she gained strength to accompany them. The wretches who had abandoned the poor creature, were on their way to a fishing station, which they knew to be very productive, and not above a day's march distant. She was unable to keep the pace at which they chose to proceed, and having no near relation but an aged aunt, who could not assist her, they left her at an encampment without any food. The hunters met this party of Indians about a month afterwards, when they were living in abundance. The girl, by that time, had perfectly recovered her strength, and they desired that she should be restored to them, but the hunters firmly resisted their importunity, and one of them adopted her as his own child. It is singular that she was the only female of the tribe that could be called good-looking. Her Indian name was Aton-larree, which the interpreter translated, Burnt-weed. Lieutenant Back made a sketch of her, in the dress which the hunter's wife gave to her on their first meeting. When the Indians came to the fort, I took the first opportunity of their being assembled in the hall, to send for the hunters and their wives, and to reward them by a substantial present of clothing and ammunition. I also gave to them some neat steel instruments, consisting of gimblets, and other useful articles, which they were desired to preserve, and shew to other Indians, as a testimony of our approbation of their

1826.
March.

humanity. A present was also bestowed upon the girl, and then the Dog-Ribs were addressed as to their unfeeling conduct towards her. They listened quietly, and merely stated her weakness as the cause. There is little doubt but that the transactions of this day were canvassed afterwards, and it is to be hoped that the knowledge of our sentiments gaining circulation, may induce a discontinuance of their inhuman practices.

Wednesday.
22nd.

By the men who had conveyed our last packet to Fort Simpson, we received intelligence that some Chipewyans had brought information to the Athabasca and Slave Lakes, of their having seen many indications of a party of white people on the sea-coast eastward of the Coppermine River. The report stated, that they had found, in the preceding autumn, on the borders of a river near the sea-coast, a sawpit, some saws, and axes, and a store of deer's meat. There was snow on the ground, and the footsteps of the party appeared recent. We concluded from these statements, that Captain Parry had laid up his ships in the vicinity of Bathurst's Inlet, and sent hunting-parties up the river to augment his stock of provision. I therefore despatched two men with letters to Mr. M'Vicar, at Slave Lake, containing a series of questions, that the matter might be thoroughly investigated, and requested him to transmit the answers to the Admiralty. I likewise begged of him immediately to procure a party of Indians to go to the spot, and convey a letter from me to Captain Parry, in order that they might either be employed as hunters for the ships, or carry their letters to the nearest establishment for conveyance to England. Had the information reached us sooner, so that a party could have gone from Bear Lake to the point at which the ships were, and returned before the men were wanted, I should have sent to ascertain the fact. The idea of the ships being on the northern coast, the prospect of their success, and the expectation of the

L 2

eastern detachment meeting them in the summer, afforded enlivening topics of conversation for several days, and on the day the intelligence came, we celebrated its arrival with a bowl of punch. The health of Captain Parry, and his party, as well as that of Captain Beechey, was drank with enthusiasm.

Thursday,
23rd.
We obtained observations for the time, from which it appeared that the chronometer, No. 1733, generously lent to the Expedition by my friend the late Mr. Moore, had only varied its rate two hundredths of a second, since the 3rd of November. I had worn it next my skin, suspended round my neck, the whole time; and, consequently, it was not exposed to much variation of temperature.

After the middle of this month the N.W. winds gave place to a succession of easterly breezes; whenever these prevailed, we observed the terrestrial refraction was much increased: double refraction of the land was not unfrequent, and twice the mist, arising from the open water, appeared like a wall of ice. When the moon shone, halos, and occasionally paraselenæ, were visible; and towards the close of the month the coruscations of the aurora were often very brilliant.

During this month I noticed that on several occasions the magnetic needle oscillated when I approached it in a dress of water-proof cloth, although it remained stationary when others of the party examined it in their ordinary garments. The waterproof dress probably acted by exciting electricity in the body, although this opinion is rather contradicted by the fact of a fur cap, which had been rubbed by the hand until it affected the gold-leaf electrometer, producing no change in the needle, and my approach to the electrometer not causing the gold-leaf to expand.

Having failed in an attempt to make charcoal for the blacksmith's use at this place, we despatched William Duncan, and the blacksmith, to make some at Fort Norman, where birch trees are plentiful; and on the 6th of April we were glad to see them return with the first load. The carpenters had already prepared the timbers and the keel for the new boat, and we were waiting for the coals to get the iron-work forward.

On the 10th Dr. Richardson and Mr. Kendall left the fort on snow-shoes, accompanied by an Indian guide, and a man driving a dog-sledge with provisions, for the purpose of completing the survey of Great Bear Lake, which Dr. Richardson had commenced the preceding autumn. The day was remarkably warm; the blackened thermometer, exposed to the sun, rose to + 90; and we hailed with delight a complete thaw. Cheered by the prospect, a spot was cleared of snow, the keel of the boat laid down, and that there might be no delay, all the sledges we could spare were despatched to fetch the remainder of the charcoal from Fort Norman. On the following day water was dripping from the roofs, and the flies were active within the rooms. The continuance of mild weather for six days caused a rapid decay of the snow, but no spots of land became visible. The men returned with the charcoal, and from them we learned that the season was more backward here than in the vicinity of Fort Norman. In the evening of the 17th, a telescope was put up in the meridian for finding the rates of the chronometers by the transit of Arcturus. On the 19th, thirty Hare-Indians arrived with sledges, bringing their winter's collection of furs for the Hudson Bay Company, and a large supply of dried meat for us, which, with the stock already in store, put us quite at ease respecting food until the season for our departure. The party consisted mostly of young lads, who, very good-naturedly, sang and danced for our amusement all the evening. They likewise gave us specimens of the dances in use among the Loucheux, which were more graceful than their own. The tune they sung to the Medicine-dance of the Loucheux, struck

1826.
April.
6th.

Monday,
10th.

Tuesday,
11th.

Wednesday,
19th.

me as being soft and pretty. The ludicrous attitudes and grotesque figures of the dancers, as they wheeled in a circle, shaking the knives and feathers which they held between their fingers, were happily sketched by Lieutenant Back.

As the fish had withdrawn from the open water at the commencement of the fine weather, the nets were brought nearer to the house; but we did not obtain more than thirty fish daily. This diminution, however, gave us no concern, as we had plenty of meat. Shortly afterwards the trout began again to take bait, and we caught several of large size. Easterly winds prevailed this month, and they blew uninterruptedly from the 21st to the last day. A storm, on the 28th and 29th, delayed the carpenters working at the boat: the patches of ground which had for the last few days been visible, were again covered with snow, and the general aspect was bleak and wintry.

Dr. Richardson and Mr. Kendall returned on the 1st of May, and we were furnished with the following particulars of their journey. Their course, on leaving us, was first directed to the fishery in Mac Vicar's Bay, which they reached on the fourth day, and from whence, taking with them another sledge-load of provisions and an additional attendant, they continued their journey to the bottom of Mac Tavish Bay, the most easterly part of the Lake. The reduction in their stock of provisions now caused them to commence their return, and they reached the fort after an absence of three weeks, during which, in very unfavourable weather, they travelled about three hundred and eighty miles. Dr. Richardson had sailed four hundred and eighty miles through the lake in the autumn, and in the two excursions, five hundred miles of its shores were delineated, and the positions of many points established by astronomical observations. About twenty miles of the north shore of Mac Tavish Bay are the only parts of Bear Lake remaining unsurveyed. The following brief descrip-

1826.
May.
Monday,
1st.

tion of Bear Lake is extracted from Dr. Richardson's Journal:—

" Great Bear Lake is formed by the union of five arms or bays, which were named after Messrs. Keith, Smith, Dease, Mac Tavish, and Mac Vicar, of the Hudson's Bay Company. The principal feeding-stream, named Dease River, rises in the Copper Mountains, and falls into the upper end of Dease Bay, which is the most northern part of the lake, and Bear Lake River, which conveys the waters of the lake to the Mackenzie, issues from Keith Bay, the most southerly arm. Mac Tavish Bay is the most easterly portion of the lake, and Smith Bay, which lies opposite to it, runs to the westward. Mac Vicar Bay has a southerly direction nearly parallel to Keith Bay. The length of the lake, from Dease River to Bear Lake River, is about one hundred and seventy-five miles; and its breadth, from the bottom of Smith Bay to the bottom of Mac Tavish Bay, is one hundred and fifty miles. A range of granite hills skirts the bottom of Mac Tavish Bay. The Great Bear Mountain, at whose base some bituminous shale cliffs are exposed, is about nine hundred feet high, and separates Mac Vicar and Keith Bays; a similar mountain lies betwixt Keith and Smith Bays. In Dease Bay, limestone and sandstone are the prevailing rocks. The waters of the lake are very clear, and of unknown depth; forty-five fathoms of line were let down near the shore, in Mac Tavish Bay, without reaching the bottom. There is a considerable quantity of good wood, principally white spruce, in the vicinity of the lake; but there is reason to believe that, before many years elapse, it will become scarce, for it is very slow of growth, and the natives every year set fire to it in various quarters, and thus destroy it for many miles. The finest timber was observed on the west side of Great Bear Lake Mountain. There are good fisheries in Dease Bay, and in various other quarters of the lake; but the fish taken in Mac Vicar Bay are remarkably fine and abundant at all seasons of the year. The principal advantage of the site chosen for Fort Franklin, is its vicinity to the Bear Lake River, and the great quantity of fish that can be procured at certain seasons, although they are small and of inferior quality."

On the 5th of this month, the men being called in from the fishery in Mac Vicar Bay, the whole party was once more assembled at the house, anxiously looking forward to the arrival of spring. We hailed the appearance of swans, on the following day, as a sure sign of its approach. A goose was seen on the 7th, two ducks on the 8th, and on the 9th several gulls were observed in the open water near the Bear Lake River. The snow, at this time, was rapidly diminishing from the surface of the lake, and there were many spots of ground visible. We, therefore, commenced the preparations for the summer's voyage. The seamen were employed in repairing the coverings and sails of the boats, as well as in refitting their rigging, and occupation was allotted to every person in the establishment. These operations requiring the constant superintendence of the officers, the observations on the magnetic-needle were discontinued. After the middle of the month, we were visited by occasional showers of rain, which removed the snow and produced a perceptible decay of the ice.

On the 23rd, the ice broke away from the shore of the small lake, and also of Bear Lake, in front of the house. Swans and geese were now daily passing to the northward; many shots were fired at them, both by the Indians and our own party, but only a few were killed. The geese were principally of the kind known to naturalists by the name of Canada geese, and denominated bustards by the voyagers. Numbers of white geese also passed; we saw only two flocks of laughing-geese. The first swallow came on the 16th, and, on the following day, many others arrived. A variety of ducks, gulls, and many of the small aquatic birds, now frequented the marshy borders of the little lake, which afforded constant amusement to the sportsmen, and full occupation to Dr. Richardson in preparing the skins for specimens.

On the 24th, the musquitoes appeared, feeble at first, but, after a few days, they became vigorous and tormenting. The first flower, a tussilago, was gathered on the 27th. Before the close of the month, several others were in bloom, of which the most abundant was the white anemone (*anemone tenella*). The leaf-buds had not yet burst, though just ready to open.

The carpenters had now finished the new boat, which received the name of the Reliance. It was constructed of fir, with birch timbers, after the model of our largest boat, the Lion, but with a more full bow, and a finer run abaft. Its length was twenty-six feet, and breadth five feet eight inches. It was fastened in the same manner as the other boats, but with iron instead of copper, and to procure sufficient nails we were obliged to cut up all the spare axes, trenches *, and ice-chisels. Being without tar, we substituted strips of water-proof canvass, soaked in some caoutchouc varnish, which we had brought out, to lay between the seams of the planks; and for paint, we made use of resin, procured from the pine-trees, boiled and mixed with grease. The other boats were afterwards put in complete repair. The Lion required the most, in consequence of the accident in Bear Lake River. The defects in the other two principally arose from their having been repaired at Cumberland House with the elm that grows in its vicinity, and is very spongy. We now substituted white spruce fir, which, when grown in these

* Used by the Indians to break up the beaver lodges.

M

high latitudes, is an excellent wood for boat-building. We were surprised to find, that, notwithstanding the many heavy blows these boats had received in their passage to this place, there was not a timber that required to be changed.

In our bustle, we would gladly have dispensed with the presence of the Dog-Ribs, who now visited us in great numbers, without bringing any supplies. They continued hanging about the fort, and their daily drumming and singing over the sick, the squalling of the children, and bawling of the men and women, proved no small annoyance. We were pleased, however, at perceiving that the ammunition we had given to them in return for meat, had enabled them to provide themselves with leathern tents. Their only shelter from wind, snow, or rain, before this season, had been a rude barricade of pine branches. Fortunately, for our comfort, they were obliged to remove before the expiration of the month to a distant fishery to procure provision.

The preparations for the voyage along the coast being now in a state of forwardness, my attention was directed to the providing for the return of Dr. Richardson's party to this establishment in the following autumn, and to the securing means of support for all the members of the Expedition at this place, in the event of the western party being likewise compelled to return to it. Respecting the first point, it was arranged that Beaulieu the interpreter, and four Canadians, should quit Fort Franklin on the 6th of August, and proceed direct to Dease River with a bateau, and wait there until the 20th of September, when, if Dr. Richardson did not appear, they were to come back to the fort in canoes, and to leave the boat, with provision and other necessaries, for the use of the eastern detachment. All these points were explained to Beaulieu, and he not only understood every part of the arrangement, but seemed very-desirous to perform the important duty entrusted to him. I next drew up written instructions for the guidance of Mr. Dease, during the absence of the Expedition, directing his attention first to the equipment and despatch of Beaulieu on the 6th of August, and then to the keeping the establishment well-stored with provision. He was aware of the probability that the western party would meet his Majesty's ship Blossom, and go to Canton in her. But as unforeseen circumstances might compel us to winter on the coast, I considered it necessary to warn him against inferring, from our not returning in the following autumn, that we had reached the Blossom. He was, therefore, directed to keep Fort Franklin complete, as to provision, until the spring of 1828. Dr. Richardson was likewise instructed, before he left the fort in 1827, on his return to England, to see that Mr. Dease fully understood my motives for giving these orders, and that he was provided with the means of purchasing the necessary provision from the Indians.

The long reign of the east wind was at length terminated by a fresh N.W. breeze, and the ice yet remaining on the small lake soon disappeared, under the softening effects of this wind. This lake had been frozen eight months, wanting three days. A narrow channel being opened along the western border of Bear Lake, on the 14th Dr. Richardson took advantage of it, and went in a small canoe with two men to examine the mountains on the borders of Bear Lake River, and to collect specimens of the plants that were now in flower, intending to rejoin the party at Fort Norman. On the same day, in 1821, the former Expedition left Fort Enterprize for the sea.

The equipments of the boats being now complete, they were launched on the small lake, and tried under oars and sails. In the afternoon the men were appointed to their respective stations, and furnished with the sky-blue waterproof uniforms, and

M 2

feathers, as well as with the warm clothing which had been provided for the voyage. I acquainted them fully with the object of the Expedition, and pointed out their various duties. They received these communications with satisfaction, were delighted with the prospect of the voyage, and expressed their readiness to commence it immediately. Fourteen men, including Augustus, were appointed to accompany myself and Lieutenant Back, in the Lion and Reliance, the two larger boats; and ten, including Ooligbuck, to go with Dr. Richardson and Mr. Kendall, in the Dolphin and Union. In order to make up the complement of fourteen for the western party, I proposed to receive two volunteers from the Canadian voyagers; and to the credit of Canadian enterprise, every man came forward. I chose François Felix and Alexis Vivier, because they were the first who offered their services, and this too without any stipulation as to increase of wages.

Spare blankets, and every thing that could be useful for the voyage, or as presents to the Esquimaux, which our stores could furnish, were divided between the eastern and western parties, and put up into bales of a size convenient for stowage. This interesting day was closed by the consumption of a small quantity of rum, reserved for the occasion, followed by a merry dance, in which all joined with great glee, in their working dresses. On the following Sunday the officers and men assembled at Divine service, dressed in their new uniforms; and in addition to the ordinary service of the day, the special protection of Providence was implored on the enterprise we were about to commence. The guns were cleaned the next day, and stowed in the arm chests, which had been made to fit the boats. Tuesday and Wednesday were set apart for the officers and men to pack their own things. A strong western breeze occurred on the 21st, which removed the ice from the front of the house and opened a passage to the Bear Lake River. The men were sent with the boats and stores to the river in the evening, and were heartily cheered on quitting the beach. The officers remained to pack up the charts, drawings, and other documents, which were to be left at the fort; and, in the event of none of the officers returning, Mr. Dease was directed to forward them to England. We quitted the house at half past ten, on Tuesday morning, leaving Cotè, the fisherman, in charge, until Mr. Dease should return from Fort Norman. This worthy old man, sharing the enthusiasm that animated the whole party, would not allow us to depart without giving his hearty, though solitary cheer, which we returned in full chorus.

The position of Fort Franklin was determined to be in latitude 65° 11′ 56″ N., longitude 123° 12′ 44″ W.; variation of the compass 39° 9′ E.; dip of the needle 82° 58′ 15″.

CHAPTER III.

Voyage to the Sea—Part from the Eastern Detachment at Point Separation—Reach the Mouth of the Mackenzie—Interview and Contest with the Esquimaux—Detained by Ice—Meet friendly Esquimaux—Point Sabine.

1826.
June.
Thursday,
22nd.

ON our arrival at the Bear Lake River, we were mortified to find the ice drifting down in large masses, with such rapidity as to render embarkation unsafe. The same cause detained us the whole of the following day; and as we had brought no more provision from the house than sufficient for an uninterrupted passage to Fort Norman, we sent for a supply of fish. This was a very sultry day, the thermometer in the shade being 71° at noon, and 74° at three P.M.

The descent of the ice having ceased at eight in the morning of the 24th, we embarked. The heavy stores were put into a bateau, manned by Canadians, who were experienced in the passage through rapids, and the rest of the boats were ordered to follow in its wake, keeping at such a distance from each other as to allow of any evolution that might be necessary to avoid the stones. The boats struck several times, but received no injury. At the foot of the rapid we met a canoe, manned by four of our Canadian voyagers, whom Dr. Richardson had sent with some letters that had arrived at Fort Norman from the Athabasca Lake; and as the services of the men were wanted, they were embarked in the boats, and the canoe was left. Shortly afterwards we overtook Beaulieu, who had just killed a young moose deer, which afforded the party two substantial meals. At this spot, and generally along the river, we found abundance of wild onions.

We entered the Mackenzie River at eight in the evening, and the current being too strong for us to advance against the stream with oars, we had recourse to the tracking line, and travelled all night. It was fatiguing, owing to large portions of the banks having been overthrown by the disruption of the ice, and from the ground being so soft that the men dragging the rope sank up to the knees at every step; but these impediments were less regarded than the ceaseless torment of the musquitoes. We halted to sup at the spot where Sir A. Mackenzie saw the flame rising from the bank in 1789. The precipice was still on fire, the smoke issuing through several apertures. Specimens of the coal were procured.

We reached Fort Norman at noon on the 25th. On the following morning the provision and stores which had been left at this place were examined, and found to be in excellent order, except the powder in one of the magazines, which had become caked from damp. I had ordered a supply of iron-work, knives, and beads, for the sea voyage from Fort Simpson; they had arrived some days before us, and with our stock thus augmented, we were well furnished with presents for the natives. The packages being finished on the 27th, the boats received their respective ladings, and we were rejoiced to find that each stowed her cargo well, and with her crew embarked floated as buoyantly as our most sanguine wishes had anticipated. The heavy stores, however, were afterwards removed into a bateau that was to be taken to the mouth of the river, to prevent the smaller boats from receiving injury in passing over the shoals.

We waited one day to make some pounded meat we had brought into pemmican. In the meantime the seamen enlarged the foresail of the Reliance.

The letters which I received from the Athabasca department informed me that the things I had required from the Company in February last, would be duly forwarded; they likewise contained a very different version of the story which had led us to suppose that Captain Parry was passing the winter on the northern coast. We now learned that the Indians had only seen some pieces of wood recently cut, and a deer that had been killed by an arrow; these things we concluded were done by the Esquimaux. Three men from Slave Lake, whom I had sent for to supply the place of our Chipewyan hunters, who were very inactive last winter, joined us at this place. They were to accompany Mr. Dease and the Canadians to Fort Franklin; and that they, as well as the Indians, might have every encouragement to exert themselves in procuring provision during the summer, I directed a supply of the goods they were likely to require, to be sent from Fort Simpson, as soon as possible. The longitude of Fort Norman was observed to be 124° 44' 47" W., its latitude 64° 40' 38" N.; variation 39° 57' 52" E.

Wednesday, 28th.
Early this morning the boats were laden and decorated with their ensigns and pendants, and after breakfast we quitted the fort, amidst the hearty cheers of our friends Mr. Dease, Mr. Brisbois, and the Canadians, and I am sure carried their best wishes for our success. We halted at noon to obtain the latitude, which placed the entrance of Bear Lake River in 64° 55' 37" N.; and Dr. Richardson took advantage of this delay to visit the mountain at that point, but his stay was short, in consequence of a favourable breeze springing up. We perceived that the four boats sailed at nearly an equal rate in light breezes, but that in strong winds the two larger ones had the advantage. When we

1826.
June.

landed to sup the musquitoes beset us so furiously that we hastily despatched the meal and re-embarked, to drive under easy sail before the current. They continued, however, to pursue us, and deprived us of all rest. On our arrival, next morning, at the place of the first rapids, there was scarcely any appearance of broken water, and the sand-bank on which Augustus had been so perilously situated in the preceding autumn, was entirely covered. This was, of course, to be ascribed to the spring floods; the increase of water to produce such a change, must have exceeded six feet. In the afternoon we were overtaken by a violent thunder-storm, with heavy rain, which made us apprehensive for the pemmican, that spoils on being wet. It unfortunately happened that a convenient place for spreading out the bags that were injured could not be found, until we reached the Hare-Skin River, below the Rampart Defile, which was at nine o'clock. They were spread out the next morning, with the other perishable parts of the cargo, and we remained until they were dry. We embarked at ten, and, aided by a favourable breeze, made good progress until six P.M., when the threatening appearance of the clouds induced us to put on shore, and we had but just covered the baggage before heavy rain fell, that continued throughout the night. Four Hare Indians came to the encampment, to whom dried meat and ammunition were given, as they were in want of food from being unable to set their nets in the present high state of the water. These were the only natives seen since our departure from Fort Norman; they informed us, that, in consequence of not being able to procure a sufficiency of fish in the Mackenzie at this season, their companions had withdrawn to gain their subsistence from the small lakes in the interior.

Friday,
30th.

We embarked at half past one on the morning of the 1st of July. The sultry weather of the preceding day made us now

July 1st.

feel more keenly the chill of a strong western breeze, and the mist which it brought on, about four hours after our departure. This wind, being contrary to the current, soon raised such high waves that the boats took in a great deal of water; and as we made but little progress, and were very cold, we landed to kindle a fire, and prepare breakfast; after which we continued the voyage to Fort Good Hope, without any of the interruptions from sand-banks that we had experienced in the autumn.

On our arrival we were saluted with a discharge of musketry by a large party of Loucheux, who had been some time waiting at the fort, with their wives and families, for the purpose of seeing us. After a short conference with Mr. Bell, the master of the post, we were informed that these Indians had lately met a numerous party of Esquimaux at the Red River, by appointment, to purchase their furs; and that in consequence of a misunderstanding respecting some bargain, a quarrel had ensued between them, which fortunately terminated without bloodshed. We could not, however, gain any satisfactory account of the movements of the Esquimaux. The only answers to our repeated questions on these points were, that the Esquimaux came in sixty canoes to Red River, and that they supposed them to have gone down the eastern channel, for the purpose of fishing near its mouth. The chief, however, informed us that he had mentioned our coming to their lands this spring, and that they had received the intelligence without comment; but from his not having alluded to this communication until the question was pressed upon him, and from the manner of his answering our inquiries, I thought it doubtful whether such a communication had really been made.

We had been led to expect much information from the Loucheux respecting the channels of the river, and the coast on the east and west side near its mouth, but we were greatly disappointed. They were ignorant of the channel we ought to follow in order to arrive at the western mouth of the river; and the only intelligence they gave us respecting the coast on that side was, that the Esquimaux represented it to be almost constantly beset by ice. They said also that they were unacquainted with the tribes who reside to the westward. Several of the party had been down the eastern channel, of which they made a rude sketch; and their account of the coast on that side was, that, as far as they were acquainted with it, it was free from ice during the summer.

Mr. C. Dease, the former master of Fort Good Hope, had retained two of the Loucheux to accompany the Expedition until we should meet the Esquimaux: they spoke a few words of the language, which they had learned during an occasional residence with the tribe that resides on the eastern border of the river. But the knowledge of the recent transactions at Red River had convinced us that their presence would be more likely to irritate than pacify the Esquimaux. We also discovered that their sole motive for accompanying us was the desire of trading with that people; and further, that they expected we should take their families and baggage in the boats. Their services were therefore declined; and a compensation was offered to them for their loss of time in waiting for us; but having fixed their minds on the gain to be derived from us and from the trade with the Esquimaux, they expressed great disappointment, and were very intemperate in their language. As I was anxious, for the sake of the trade at the post, to leave them no room to complain either of us, or of Mr. Dease who had acted for us, I spent several hours in debate with them to very little purpose, and at last discovered that the whole scene was got up for the purpose of obtaining a few more goods. My compliance with their wish rendered them quite contented. I afterwards

N 2

added a present to the principal chief of the party, who still expressed a wish to accompany us, but he frankly said that if he went, all his young men must go also. They came in the evening in great good humour to exhibit their dances in front of our tent, a compliment we could well have dispensed with, as we were busy.

Having ascertained that the Esquimaux were likely to be seen in greater numbers than had been at first imagined, I increased the stock of presents from the store at this place, and exchanged two of our guns, which were defective, that the party might have entire confidence in their arms. And to provide against the casualty of either or both branches of the Expedition having to return this way, I requested Mr. Bell to store up as much meat as he could during the summer. We learned from this gentleman that the supply of meat at this post was very precarious, and that had we not left the five bags of pemmican in the autumn, the residents would have been reduced to great distress for food during the winter. These bags were now replaced. The arrangements being concluded, we spent the greater part of the night in writing to England. I addressed to the Colonial Secretary an account of our proceedings up to this time, and I felt happy to be able to state that we were equipped with every requisite for the Expedition.

Sunday, 2nd.

We quitted Fort Good Hope at five on the 2nd. In the passage down the river we were visited by several Loucheux, who, the instant we appeared, launched their canoes, and came off to welcome us. We landed, at their request, to purchase fish; yet, after the bargain had been completed, an old woman stepped forward, and would only allow of our receiving two fish: she maintained her point, and carried off the rest in spite of all remonstrance. The natives were all clothed in new leathern dresses, and looked much neater, and in better health, than last autumn.

Being anxious to reach the Red River, we continued rowing against the wind until after midnight. On reaching that place, the ground proved too wet for us to encamp; we, therefore, proceeded a short distance lower down, and put up under some sandstone cliffs, where there was but just room for the tents. As we were now on the borders of the Esquimaux territory, we devoted the following morning to cleaning the arms; and a gun, dagger, and ammunition, were issued to each person. We had no reason, indeed, to apprehend hostility from the Esquimaux, after the messages they had sent to Fort Franklin, but vigilance and precaution are never to be omitted in intercourse with strange tribes.

Embarking at two in the afternoon of the 3rd, we soon entered the expansion of the river whence the different channels branch off, and steering along the western shore, we came to the head of a branch that flowed towards the Rocky Mountain range. Being anxious not to take the eastern detachment out of their course, I immediately encamped to make the necessary arrangements for the separation of the parties. The warm clothing, shoes, and articles for presents, had been previously put up in separate packages, but the provisions remained to be divided, which was done in due proportion. Twenty-six bags of pemmican, and two of grease, were set apart for the Dolphin and Union, with a supply of arrow-root, macaroni, flour, and portable soup, making in all eighty days' provision, with an allowance for waste. The Lion and Reliance received thirty-two bags of pemmican, and two of grease, with sufficient arrow-root, &c., to make their supply proportionate to that of the eastern party. Provided no accident occurred, neither party could be in absolute want for the whole summer, because at two-thirds allowance the pemmican could be made to last one hundred days; and we had reason to expect to meet with deer occasionally.

In the evening I delivered my instructions to Dr. Richardson; they were in substance as follows:—He was to take under his charge Mr. E. N. Kendall, and ten men, and proceed in the Dolphin and Union to survey the coast between the Mackenzie and Copper-Mine Rivers. On reaching the latter river, he was to travel by land to the northeast arm of Great Bear Lake, where Beaulieu was under orders to meet him with a boat for the conveyance of his party to Fort Franklin. But if he should be so much delayed on the coast as to have no prospect of reaching the Copper-Mine River by the close of August, or the Bear Lake Portage by the 20th of September, he was not to expose himself or his party to risk by persevering beyond the 15th or 20th of August, but was to return to Fort Franklin by the way of the Mackenzie, or by any other route he might discover. The only cause of regret I had respecting the equipment of the eastern party was my being unable to provide Dr. Richardson with a chronometer, the main-springs of two out of the three chronometers furnished to us having been broken. I borrowed, however, from Mr. Dease, a watch, made by Barraud, to enable Mr. Kendall to obtain the longitude by lunar distances. They were likewise provided with that excellent instrument Massey's Log; and knowing Mr. Kendall's intimate acquaintance with marine surveying, I had no doubt of his being able to make a correct survey of the coast. The spot where the above arrangements were made, bears the name of Point Separation, and lies in latitude 67° 38' N., longitude 133° 53' W.

As the parties entertained for each other sentiments of true friendship and regard, it will easily be imagined that the evening preceding our separation was spent in the most cordial and cheerful manner. We felt that we were only separating to be employed on services of equal interest; and we looked forward with delight to our next meeting, when, after a successful

1826.
July.

termination, we might recount the incidents of our respective voyages. The best supper our means afforded was provided, and a bowl of punch crowned the parting feast.

We were joined by an elderly Loucheux, who gave us a better account of the eastern and western channels than we had hitherto obtained. " The west branch," he said, " would take us to the sea, and flowed the whole way at no great distance from the mountains." " The eastern was a good channel, and passed close to the hills on that side." He further informed us that the Esquimaux were generally to be found on an island in the eastern channel, but were seldom seen in the western branch. He was, however, unacquainted with the coast, and we found afterwards that he knew little about the movements of the Esquimaux.

By six in the morning of the 4th the boats were all laden, and ready for departure. It was impossible not to be struck with the difference between our present complete state of equipment and that on which we had embarked on our former disastrous voyage. Instead of a frail bark canoe, and a scanty supply of food, we were now about to commence the sea voyage in excellent boats, stored with three months' provision. At Dr. Richardson's desire the western party embarked first. He and his companions saluted us with three hearty cheers, which were warmly returned; and as we were passing round the point that was to hide them from our view, we perceived them also embarking. Augustus was rather melancholy, as might have been expected, on his parting from Ooligbuck, to proceed he knew not whither; but he recovered his wonted flow of spirits by the evening.

The western party were distributed as follows :—

Tuesday,
4th.

LION.	RELIANCE.
John Franklin, *Captain R. N.*	George Back, *Lieutenant R. N.*
William Duncan, *Cockswain.*	Robert Spinks, *Cockswain.*
Thomas Matthews, *Carpenter.*	Robert Hallom, *Corpl. of Marines.*
Gustavus Aird, *Bowman.*	Charles Mackenzie, *Bowman.*
George Wilson, *Marine.*	Alexander Currie, *Middle Man.*
Archibald Stewart, *Soldier.*	Robert Spence, *Ditto*
Neil Mac Donald, *Voyager.*	Alexis Vivier, *Canadian.*
Augustus, *Esquimaux.*	François Felix, *Ditto.*

Our course was directly towards the Rocky Mountain range, till we came near the low land that skirts its base; where, following the deepest channel, we turned to the northward. I was desirous of coasting the main shore, but finding some of the westernmost branches too shallow, we kept on the outside of three islands for about twelve miles, when we entered the channel that washes the west side of Simpson's Island. It was winding, and its breadth seldom exceeded a quarter of a mile. During our progress we occasionally caught a glimpse of the Rocky Mountains, which was an agreeable relief to the very dull picture that the muddy islands in our neighbourhood afforded. We halted to breakfast just before noon, and observed the latitude 67° 51′ N.

In the afternoon one deer was seen, and many swans and geese; we did not fire at them, for fear of alarming any Esquimaux that might be near. Encamped at eight P.M., opposite Simpson's Island, in latitude 68° 13′ N., longitude 134° 27′ W. The boats were secured without discharging the cargoes, and two men were placed on guard, to be relieved every two hours.

Wednesday, 5th.

We set forward at four A.M., with a favourable breeze, and made good progress, though the river was very winding. At eight we entered a branch that turned to the westward round the point of Halkett Island into the channel washing the main

1826.
July.

shore.　We soon afterwards arrived at a spot where a large body of Esquimaux had been encamped in the spring, and supposing that they might revisit this place, a present of an ice-chisel, kettle, and knife, was hung up in a conspicuous situation.　Soon after we had entered the channel that flows by the main shore, we first perceived lop-sticks, or pine trees, divested of their lower branches, for the purpose of land-marks, and therefore concluded it was much frequented by the Esquimaux.　Our course was then altered to N.W., and we soon passed the last of the well-wooded islands.　The spruce fir-trees terminated in latitude 68° 36′ N.; and dwarf willows only grew below this part.　A very picturesque view was obtained of the Rocky Mountains, and we saw the entire outline of their peaked hills, table-land, and quoin-shaped terminations.　Two lofty ranges were fronted by a lower line of round-backed hills, in which we perceived the strata to be horizontal, and the stone of a yellow colour.　A few miles lower down we found hills of sand close to the west border of the river.　We passed several deserted huts, and in one spot saw many chips and pieces of split drift-wood, that appeared to have been recently cut.　The channel varied in breadth from a half to three-quarters of a mile, but, except in the stream of the current, the water was so shallow as scarcely to float the boats, and its greatest depth did not exceed five feet.　We landed at eight P.M., on Halkett Island, intending to encamp, but owing to the swampiness of the ground the tent could not be pitched. Having made a fire and cooked our supper, we retired to sleep under the coverings of the boats, which afforded us good shelter from a gale and heavy rain that came on before midnight.　Latitude 68° 39′ N., longitude 135° 35′ W.

Thursday,
6th.

The continuance of stormy weather detained us until two P.M. of the 6th, when, the rain ceasing, we embarked.　After passing through the expansion of the river near the west extreme of Hal-

kett Island, we turned into the narrower and more winding channel, between Colvill Island and the main. A fog coming on at
eight P.M. we encamped, in latitude 68° 48′ N., longitude 136° 4′
W.; temperature of the air 42°, that of the water being 47° in
the middle of the stream. Several of the glaucous gulls were seen,
and this circumstance, as well as a line of bright cloud to the
N.W. resembling the ice-blink, convinced us that the sea was
not far off. A rein-deer appearing near the encampment, two
men were sent after it, who returned unsuccessful. Augustus
obtained a goose for supper. Many geese, swans, and ducks, had
been seen on the marshy shores of the island in the course of
the day.

The night was cold, and at day-light on the 7th the thermometer indicated 36°. Embarking at four A.M. we sailed down
the river for two hours, when our progress was arrested by the
shallowness of the water. Having endeavoured, without effect,
to drag the boats over the flat, we remounted the stream to
examine an opening to the westward, which we had passed. On
reaching the opening we found the current setting through it
into the Mackenzie, by which we knew that it could not afford a
passage to the sea, but we pulled up it a little way, in the hope
of obtaining a view over the surrounding low grounds from the
top of an Esquimaux house which we saw before us. A low
fog, which had prevailed all the morning, cleared away, and we
discovered that the stream we had now ascended issued from a
chain of lakes lying betwixt us and the western hills, which were
about six miles distant, the whole intervening country between
the hills and the Mackenzie being flat.

After obtaining an observation for longitude in 136° 19′ W.,
and taking the bearing of several remarkable points of the Rocky
Mountain range, we returned to the Mackenzie, and passing the
shallows which had before impeded us, by taking only half the

boats' cargoes over at a time, we came in sight of the mouth of the river. Whilst the crews were stowing the boats, I obtained an observation for latitude in 68° 53′ N., and having walked towards the mouth of the river, discovered on an island, which formed the east side of the bay into which the river opened, a crowd of tents, with many Esquimaux strolling amongst them. I instantly hastened to the boats, to make preparations for opening a communication with them, agreeably to my instructions. A selection of articles for presents and trade being made, the rest of the lading was closely covered up; the arms were inspected, and every man was directed to keep his gun ready for immediate use. I had previously informed Lieutenant Back of my intention of opening the communication with the Esquimaux by landing amongst them, accompanied only by Augustus; and I now instructed him to keep the boats afloat, and the crews with their arms ready to support us in the event of the natives proving hostile; but on no account to fire until he was convinced that our safety could be secured in no other way. Having received an impression from the narratives of different navigators that the sacrifices of life which had occurred in their interviews with savages, had been generally occasioned by the crews mistaking noise and violent gestures for decided hostility, I thought it necessary to explain my sentiments on this point to all the men, and peremptorily forbade their firing till I set the example, or till they were ordered to do so by Lieutenant Back. They were also forbidden to trade with the natives on any pretence, and were ordered to leave every thing of that kind to the officers.

On quitting the channel of the river we entered into the bay, which was about six miles wide, with an unbounded prospect to seaward, and steered towards the tents under easy sail, with the ensigns flying. The water became shallow as we drew towards the island, and the boats touched the ground when about a mile from the beach; we shouted, and made signs to the Esquimaux to come off, and then pulled a short way back to await their arrival in deeper water. Three canoes instantly put off from the shore, and before they could reach us others were launched in such quick succession, that the whole space between the island and the boats was covered by them. The Esquimaux canoes contain only one person, and are named *kaiyacks;* but they have a kind of open boat capable of holding six or eight people, which is named *oomiak.* The men alone use the kaiyacks, and the oomiaks are allotted to the women and children. We endeavoured to count their numbers as they approached, and had proceeded as far as seventy-three canoes, and five oomiaks, when the sea became so crowded by fresh arrivals, that we could advance no farther in our reckoning. The three headmost canoes were paddled by elderly men, who, most probably, had been selected to open the communication. They advanced towards us with much caution, halting when just within speaking distance, until they had been assured of our friendship, and repeatedly invited by Augustus to approach and receive the present which I offered to them Augustus next explained to them in detail the purport of our visit, and told them that if we succeeded in finding a navigable channel for large ships, a trade highly beneficial to them would be opened. They were delighted with this intelligence, and repeated it to their countrymen, who testified their joy by tossing their hands aloft, and raising the most deafening shout of applause I ever heard.

. After the first present, I resolved to bestow no more gratuitously, but always to exact something, however small, in return; the three elderly men readily offered the ornaments they wore in their cheeks, their arms, and knives, in exchange for the articles I gave them. Up to this time the first three were the only kaiyacks that had ventured near the boats, but the natives around us had now increased to two hundred and fifty or three hundred persons, and they all became anxious to share in the lucrative trade which they saw established, and pressed eagerly upon us, offering for sale their bows, arrows, and spears, which they had hitherto kept concealed within their canoes. I endeavoured in vain, amidst the clamour and bustle of trade, to obtain some information respecting the coast, but finding the natives becoming more and more importunate and troublesome, I determined to leave them, and, therefore, directed the boats' heads to be put to seaward. Notwithstanding the forwardness of the Esquimaux, which we attributed solely to the desire of a rude people to obtain the novel articles they saw in our possession, they had hitherto shown no unfriendly disposition; and when we told them of our intention of going to sea, they expressed no desire to detain us, but, on the contrary, when the Lion grounded in the act of turning, they assisted us in the kindest manner by dragging her round. This manœuvre was not of much advantage to us, for, from the rapid ebbing of the tide, both boats lay aground; and the Esquimaux told us, through the medium of Augustus, that the whole bay was alike flat, which we afterwards found to be correct.

An accident happened at this time, which was productive of unforeseen and very annoying consequences. A kaiyack being overset by one of the Lion's oars, its owner was plunged into the water with his head in the mud, and apparently in danger of being drowned. We instantly extricated him from his unpleasant situation, and took him into the boat until the water could be thrown out of his kaiyack, and Augustus, seeing him shivering with cold, wrapped him up in his own great coat. At first he was exceedingly angry, but soon became reconciled to his situation, and looking about, discovered that we had many bales, and other articles, in the boat, which had been concealed from the people in the kaiyacks, by the coverings being carefully spread over all. He soon began to ask for every thing he saw, and expressed much displeasure on our refusing to comply with his demands; he also, as we afterwards learned, excited the cupidity of others by his account of the inexhaustible riches in the Lion, and several of the younger men endeavoured to get into both our boats, but we resisted all their attempts. Though we had not hitherto observed any of them stealing, yet they showed so much desire to obtain my flag, that I had it furled and put out of sight, as well as every thing else that I thought could prove a temptation to them. They continued, however, to press upon us so closely, and made so many efforts to get into the boats, that I accepted the offer of two chiefs, who said that if they were allowed to come in, they would keep the others out. For a time they kept their word, and the crews took advantage of the respite thus afforded, to endeavour to force the boats towards the river into deeper water. The Reliance floated, but the Lion was immovable, and Lieutenant Back dropping astern again made his boat fast to the Lion by a rope. At this time one of the Lion's crew perceived that the man whose kaiyack had been upset had a pistol under his shirt, and was about to take it from him, but I ordered him to desist, as I thought it might have been purchased from the Loucheux. It had been, in fact, stolen from Lieutenant Back, and the thief, perceiving our attention directed to it, leaped out of the boat, and joined his countrymen, carrying with him the great coat which Augustus had lent him.

The water had now ebbed so far, that it was not knee deep at the boats, and the younger men wading in crowds around us, tried to steal every thing within their reach; slyly, however, and with so much dexterity, as almost to escape detection. The moment this disposition was manifested, I directed the crews not to suffer any one to come alongside, and desired Augustus to tell the two chiefs, who still remained seated in the Lion, that the noise and confusion occasioned by the crowd around the boats greatly impeded our exertions; and that if they would go on shore and leave us for the present, we would hereafter return from the ship which we expected to meet near this part of the coast, with a more abundant supply of goods. They received this communication with much apparent satisfaction, and jumping out of the boats repeated the speech aloud to their companions. From the general exclamation of "*teyma,*" which followed, and from perceiving many of the elderly men retire to a distance, I conceived that they acquiesced in the propriety of the suggestion, and that they were going away, but I was much deceived. They only retired to concert a plan of attack, and returned in a short time shouting some words which Augustus could not make out. We soon, however, discovered their purport, by two of the three chiefs who were on board the Reliance, jumping out, and, with the others who hurried to their assistance, dragging her towards the south shore of the river. Lieutenant Back desired the chief who remained with him to tell them to desist, but he replied by pointing to the beach, and repeating the word *teyma, teyma,* with a good-natured smile. He said, however, something to those who were seated in the canoes that were alongside, on which they threw their long knives and arrows into the boat, taking care, in so doing, that the handles and feathered ends were turned towards the crew, as an indication of pacific intentions.

As soon as I perceived the Reliance moving under the efforts of the natives, I directed the Lion's crew to endeavour to follow her, but our boat remained fast until the Esquimaux lent their aid and dragged her after the Reliance. Two of the most powerful men, jumping on board at the same time, seized me by the wrists and forced me to sit between them; and as I shook them loose two or three times, a third Esquimaux took his station in front to catch my arm whenever I attempted to lift my gun, or the broad dagger which hung by my side. The whole way to the shore they kept repeating the word " *teyma*," beating gently on my left breast with their hands, and pressing mine against their breasts. As we neared the beach, two oomiaks, full of women, arrived, and the " *teymas*" and vociferation were redoubled. The Reliance was first brought to the shore, and the Lion close to her a few seconds afterwards. The three men who held me now leaped ashore, and those who had remained in their canoes taking them out of the water, carried them to a little distance. A numerous party then drawing their knives, and stripping themselves to the waist, ran to the Reliance, and having first hauled her as far up as they could, began a regular pillage, handing the articles to the women, who, ranged in a row behind, quickly conveyed them out of sight. Lieutenant Back and his crew strenuously, but good-humouredly, resisted the attack, and rescued many things from their grasp, but they were overpowered by numbers, and had even some difficulty in preserving their arms. One fellow had the audacity to snatch Vivier's knife from his breast, and to cut the buttons from his coat, whilst three stout Esquimaux surrounded Lieutenant Back with up-lifted daggers, and were incessant in their demands for whatever attracted their attention, especially for the anchor buttons which he wore on his waistcoat. In this juncture a young chief coming to his aid, drove the assailants away. In their retreat they carried off a writing desk and cloak, which the chief rescued, and then seating himself on Lieutenant Back's knee, he endeavoured to persuade his countrymen to desist by vociferating " *teyma teyma*," and was, indeed, very active in saving whatever he could from their depredations. The Lion had hitherto been beset by smaller numbers, and her crew, by firmly keeping their seats on the cover spread over the cargo, and by beating the natives off with the butt-ends of their muskets, had been able to prevent any article of importance from being carried away. But as soon as I perceived that the work of plunder was going on so actively in the Reliance, I went with Augustus to assist in repressing the tumult; and our bold and active little interpreter rushed among the crowd on shore, and harangued them on their treacherous conduct, until he was actually hoarse. In a short time, however, I was summoned back by Duncan, who called out to me that the Esquimaux had now commenced in earnest to plunder the Lion, and, on my return, I found the sides of the boat lined with men as thick as they could stand, brandishing their knives in the most furious manner, and attempting to seize every thing that was movable; whilst another party was ranged on the outside ready to bear away the stolen goods. The Lion's crew still kept their seats, but as it was impossible for so small a number to keep off such a formidable and determined body, several articles were carried off. Our principal object was to prevent the loss of the arms, oars, or masts, or any thing on which the continuance of the voyage, or our personal safety, depended. Many attempts were made to purloin the box containing the astronomical instruments, and Duncan, after thrice rescuing it from their hands, made it fast to his leg with a cord, determined that they should drag him away also if they took it.

In the whole of this unequal contest, the self-possession of our men was not more conspicuous than the coolness with which the Esquimaux received the heavy blows dealt to them with the

P

butts of the muskets. But at length, irritated at being so often foiled in their attempts, several of them jumped on board and forcibly endeavoured to take the daggers and shot-belts that were about the men's persons; and I myself was engaged with three of them who were trying to disarm me. Lieutenant Back perceiving our situation, and fully appreciating my motives in not coming to extremities, had the kindness to send to my assistance the young chief who had protected him, and who, on his arrival, drove my antagonists out of the boat. I then saw that my crew were nearly overpowered in the fore part of the boat, and hastening to their aid, I fortunately arrived in time to prevent George Wilson from discharging the contents of his musket into the body of an Esquimaux. He had received a provocation of which I was ignorant until the next day, for the fellow had struck at him with a knife, and cut through his coat and waistcoat; and it was only after the affray was over that I learned that Gustavus Aird, the bowman of the Lion, and three of the Reliance's crew, had also narrowly escaped from being wounded, their clothes being cut by the blows made at them with knives. No sooner was the bow cleared of one set of marauders than another party commenced their operations at the stern. My gun was now the object of the struggle, which was beginning to assume a more serious complexion, when the whole of the Esquimaux suddenly fled, and hid themselves behind the drift timber and canoes on the beach. It appears that by the exertions of the crew, the Reliance was again afloat, and Lieutenant Back wisely judging that this was the proper moment for more active interference, directed his men to level their muskets, which had produced that sudden panic. The Lion happily floated soon after, and both were retiring from the beach, when the Esquimaux having recovered from their consternation, put their kaiyacks in the water, and were preparing to follow us; but I desired Augustus to say that I would shoot the first man who came within range of our muskets, which prevented them.

It was now about eight o'clock in the evening, and we had been engaged in this harassing contest for several hours, yet the only things of importance which they had carried off were the mess canteen and kettles, a tent, a bale containing blankets and shoes, one of the men's bags, and the jib-sails. The other articles they took could well be spared, and they would, in fact, have been distributed amongst them, had they remained quiet. The place to which the boats were dragged is designated by the name of Pillage Point. I cannot sufficiently praise the fortitude and obedience of both the boats' crews in abstaining from the use of their arms. In the first instance I had been influenced by the desire of preventing unnecessary bloodshed, and afterwards, when the critical situation of my party might have well warranted me in employing more decided means for their defence, I still endeavoured to temporize, being convinced that as long as the boats lay aground, and we were beset by such numbers, armed with long knives, bows, arrows, and spears, we could not use fire-arms to advantage. The howling of the women, and the clamour of the men, proved the high excitement to which they had wrought themselves; and I am still of opinion that, mingled as we were with them, the first blood we had shed would have been instantly revenged by the sacrifice of all our lives.

The preceding narrative shows that, bad as the general conduct of the Esquimaux was, we had some active friends amongst them; and I was particularly desirous of cultivating a good understanding with them, for we were as yet ignorant of the state of the ice at sea, and did not know how long we should have to remain in their neighbourhood. I was determined, however, now to keep

P 2

them at bay, and to convince them, if they made any further attempts to annoy us, that our forbearance had proceeded from good-will, and not from the want of the power to punish them. We had not gone above a quarter of a mile from Pillage Point before the boats again took the ground at the distance of one hundred and fifty yards from the shore; and having ascertained by the men wading in every direction, that there was no deeper water, we made the boats fast side by side, and remained in that situation five hours.

Shortly after the boats had been secured, seven or eight of the natives walked along the beach, and carrying on a conversation with Augustus, invited him to a conference on shore. I was at first very unwilling to permit him to go, but the brave little fellow entreated so earnestly that I would suffer him to land and reprove the Esquimaux for their conduct, that I at length consented, and the more readily, on seeing that the young chief who had acted in so friendly a manner was amongst the number on the beach. By the time that Augustus reached the shore, the number of Esquimaux amounted to forty, and we watched with great anxiety the animated conversation he carried on with them. On his return he told us that its purport was as follows:—" Your conduct," said he, " has been very bad, and unlike that of all other Esquimaux. Some of you even stole from me, your countryman, but that I do not mind; I only regret that you should have treated in this violent manner the white people who came solely to do you kindness. My tribe were in the same unhappy state in which you now are, before the white people came to Churchill, but at present they are supplied with every thing they need, and you see that I am well clothed; I get all that I want, and am very comfortable. You cannot expect, after the transactions of this day, that these people will ever bring goods to your country again, unless you show your contrition by returning the stolen goods. The white people love the Esquimaux, and wish to shew them the same kindness that they bestow upon the Indians: do not deceive yourselves, and suppose that they are afraid of you; I tell you they are not, and that it is entirely owing to their humanity that many of you were not killed to-day; for they have all guns, with which they can destroy you either when near or at a distance. I also have a gun, and can assure you that if a white man had fallen, I would have been the first to have revenged his death."

The veracity of Augustus was beyond all question with us; such a speech delivered in a circle of forty armed men, was a remarkable instance of personal courage. We could perceive, by the shouts of applause with which they filled the pauses in his harangue, that they assented to his arguments, and he told us that they had expressed great sorrow for having given us so much cause of offence, and pleaded, in mitigation of their conduct, that they had never seen white people before, that every thing in our possession was so new to them, and so desirable, that they could not resist the temptation of stealing, and begged him to assure us that they never would do the like again, for they were anxious to be on terms of friendship with us, that they might partake of the benefits which his tribe derived from their intercourse with the white people. I told Augustus to put their sincerity to the test by desiring them to bring back a large kettle and the tent, which they did, together with some shoes, having sent for them to the island whither they had been conveyed. After this act of restitution, Augustus requested to be permitted to join a dance to which they had invited him, and he was, for upwards of an hour, engaged in dancing and singing with all his might in the midst of a company who were all armed with knives, or bows and arrows. He afterwards told us that he was much delighted on finding that the words of the song, and the different attitudes of the dances, were precisely similar to those used in his own country when a friendly meeting took place with strangers. Augustus now learned from them that there was a regular ebb and flow of the tide in this bay, and that when the sun came round to a particular point there would be water enough to float the boats, if we kept along the western shore. This communication relieved me from much anxiety, for the water was perfectly fresh, and from the flood-tide having passed unperceived whilst we were engaged with the Esquimaux, it appeared to us to have been subsiding for the preceding twelve hours, which naturally excited doubts of our being able to effect a passage to the sea in this direction.

The Esquimaux gradually retired as the night advanced; and when there were only a few remaining, two of our men were sent to a fire which they had made, to prepare chocolate for the refreshment of the party. Up to this period we remained seated in the boats, with our muskets in our hands, and keeping a vigilant look out on Augustus, and the natives around him. As they had foretold, the water began to flow about midnight, and

by half past one in the morning of the 8th it was sufficiently deep to allow of our dragging the boats forward to a part where they floated. We pulled along the western shore about six miles, till the appearance of the sky bespoke the immediate approach of a gale; and we had scarcely landed before it came on with violence, and attended with so much swell as to compel us to unload the boats and drag them up on the beach.

The whole party having been exhausted by the labour and anxiety of the preceding twenty-four hours, two men were appointed to keep watch, and the rest slept until eleven o'clock in the morning, when we began to repair the damage which the sails and rigging had sustained from the attempts made by the Esquimaux to cut away the copper thimbles. We were thus employed when Lieutenant Back espied, through the haze, the whole body of the Esquimaux paddling towards us. Uncertain of the purport of their visit, and not choosing to open a conference with so large a body in a situation so disadvantageous as our present one, we hastened to launch the boats through the surf, and load them with our utmost speed; conceiving that when once fairly afloat, we could keep any number at bay. We had scarcely pulled into deep water before some of the kaiyacks had arrived within speaking distance, and the man in the headmost one, holding out a kettle, called aloud that he wished to return it, and that the oomiak which was some distance behind, contained the things that had been stolen from us, which they were desirous of restoring, and receiving in return any present that we might be disposed to give. I did not deem it prudent, however, for the sake of the few things in their possession which we required, to hazard their whole party collecting around us, and, therefore, desired Augustus to tell them to go back; but they continued to advance until I fired a ball ahead of the leading canoe, which had the desired effect—the whole party veering round, except four, who followed us for a little way, and then went back to join their companions.

I have been minute in my details of our proceedings with these Esquimaux, for the purpose of elucidating the character of the people we had to deal with; and I feel that the account would be incomplete without the mention, in this place, of some communications made to us in the month of August following, which fully explained the motives of their conduct. We learned that up to the time that the kaiyack was upset, the

Esquimaux were actuated by the most friendly feelings towards us, but that the fellow whom we had treated so kindly after the accident, discovering what the boats contained, proposed to the younger men to pillage them. This suggestion was buzzed about, and led to the conference which the old men held together when I desired them to go away, in which the robbery was decided upon, and a pretty general wish was expressed that it should be attended with the total massacre of our party. Providentially a few suggested the impropriety of including Augustus; and for a reason which could scarcely have been imagined. "If we kill him," said they, "no more white people will visit our lands, and we shall lose the opportunity of getting another supply of their valuable goods; but if we spare him, he can be sent back with a story which we shall invent to induce another party of white people to come among us." This argument prevailed at the time; but after the interviews with Augustus at the dance, they retired to their island, where they were so much inflamed by the sight of the valuable articles which they had obtained, that they all, without exception, regretted that they had allowed us to escape. While in this frame of mind the smoke of our fire being discovered, a consultation was immediately held, and a very artful plan laid for the destruction of the party, including Augustus, whom they conceived to be so firmly attached to us that it was in vain to attempt to win him to their cause. They expected to find us on shore; but to provide against the boats getting away if we should have embarked, they caused some kettles to be fastened conspicuously to the leading kaiyack, in order to induce us to stop. The kaiyacks were then to be placed in such a position as to hamper the boats, and their owners were to keep us in play until the whole party had come up, when the attack was to commence. Through the blessing of Providence, their scheme was frustrated.

But to resume the narrative of the voyage. The breeze became moderate and fair; the sails were set, and we passed along the coast in a W.N.W. direction, until eleven in the evening, when we halted on a low island, covered with drift wood, to repair the sails, and to put the boats in proper order for a sea voyage. The continuance and increase of the favourable wind urged us to make all possible despatch, and at three in the morning of the 9th again embarking, we kept in three fathoms water at the distance of two miles from the land. After sailing twelve miles, our progress was completely stopped by the ice adhering to the shore, and stretching beyond the limits of our view to seaward. We could not effect a landing until we had gone back some miles, as we had passed a sheet of ice which was fast to the shore; but at length a convenient spot being found, the boats were hauled up on the beach. We quickly ascended to the top of the bank to look around, and from thence had the mortification to perceive that we had just arrived in time to witness the first rupture of the ice. The only lane of water in the direction of our course was that from which we had been forced to retreat: in every other part the sea appeared as firmly frozen as in winter; and even close to our encampment the masses of ice were piled up to the height of thirty feet. Discouraging as was this prospect, we had the consolation to know that our store of provision was sufficiently ample to allow of a few days' detention.

The coast in this part consists of black earth, unmixed with stones of any kind, and its general elevation is from sixty to eighty feet, though in some places it swells into hills of two hundred and fifty feet. A level plain, abounding in small lakes, extends from the top of these banks to the base of a line of hills which lie in front of the Rocky Mountains. The plain was clothed with grass and plants, then in flower, specimens of which

1826.
July.

Sunday,
9th.

1826.
July.

were collected. We recognised in the nearest range of the Rocky Mountains, which I have named after my much-esteemed companion Dr. Richardson, the Fitton and Cupola Mountains, which we had seen from Garry Island at the distance of sixty miles. Few patches of snow were visible on any part of the range.

Having obtained observations for longitude and variation, we retired to bed about eight A.M., but had only just fallen asleep when we were roused by the men on guard calling out that a party of Esquimaux were close to the tents; and, on going out, we found the whole of our party under arms. Three Esquimaux had come upon us unawares, and, in terror at seeing so many strangers, they were on the point of discharging their arrows, when Augustus's voice arrested them, and by explaining the purpose of our arrival, soon calmed their fears. Lieutenant Back and I having made each of them a present, and received in return some arrows, a very amicable conference followed, which was managed by Augustus with equal tact and judgment. It was gratifying to observe our visitors jumping for joy as he pointed out the advantages to be derived from an intercourse with the white people, to whom they were now introduced for the first time. We found that they belonged to a party whose tents were pitched about two miles from us; and as they were very desirous that their friends might also enjoy the gratification of seeing us, they begged that Augustus would return with them to convey the invitation; which request was granted at his desire.

Before their departure, marks being set up on the beach one hundred and fifty yards in front of the tent, and twice that distance from the boats, they were informed that this was the nearest approach which any of their party would be permitted to make; and that at this boundary only would gifts be made, and barter carried on.   Augustus was likewise desired to explain to them the destructive power of our guns, and to assure them that every person would be shot who should pass the prescribed limit.   This plan was adopted in all succeeding interviews with the Esquimaux.   After five hours' absence Augustus returned, accompanied by twenty men and two elderly women, who halted at the boundary.   They had come without bows or arrows, by the desire of Augustus, and, following his instruction, each gave Lieutenant Back and myself a hearty shake of the hand.   We made presents to every one, of beads, fish-hooks, awls, and trinkets; and that they might have entire confidence in the whole party, our men were furnished with beads to present to them. The men were directed to advance singly, and in such a manner as to prevent the Esquimaux from counting our number, unless they paid the greatest attention, which they were not likely to do while their minds were occupied by a succession of novelties.

Our visitors were soon quite at ease, and we were preparing to question them respecting the coast, and the time of removal of the ice, when Augustus begged that he might put on his gayest dress, and his medals, before the conference began.   This was the work of a few seconds; but when he returned, surprise and delight at his altered appearance and numerous ornaments so engaged their minds, that their attention could not be drawn to any other subject for the next half hour.   "Ah," said an old man, taking up his medals, " these must have been made by such people as you have been describing, for none that we have seen could do any thing like it;" then taking hold of his coat, he asked " what kind of animal do these skins which you and the chiefs wear belong to ? we have none such in our country." The anchor buttons also excited their admiration.   At length

Q 2

we managed to gain their attention, and were informed that, as soon as the wind should blow strong from the land, the ice might be expected to remove from the shore, so as to open a passage for boats, and that it would remain in the offing until the reappearance of the stars. " Further to the westward," they continued, " the ice often adheres to the land throughout the summer ; and when it does break away, it is carried but a short distance to seaward, and is brought back whenever a strong wind blows on the coast. If there be any channels in those parts, they are unsafe for boats, as the ice is continually tossing about." " We wonder, therefore," they said, " that you are not provided with sledges and dogs, as our men are, to travel along the land, when these interruptions occur." They concluded by warning us not to stay to the westward after the stars could be seen, because the winds would then blow strong from the sea, and pack the ice on the shore. On further inquiry we learned that this party is usually employed, during the summer, in catching whales and seals, in the vicinity of the Mackenzie, and that they seldom travel to the westward beyond a few days' journey. We were, therefore, not much distressed by intelligence which we supposed might have originated in exaggerated accounts received from others. In the evening Augustus returned with them to their tents, and two of the men undertook to fetch a specimen of the rock from Mount Fitton, which was distant about twenty miles. The following observations were obtained :—Latitude 69° 1' 24" N. ; longitude 137° 35' W.; variation 46° 41' E.; dip 82° 22'.

Lieutenant Back made the sketch of the encampment from which the annexed engraving is taken. The party assembled at divine service in the evening. The wind blew in violent squalls during the night, which brought such a heavy swell upon the ice, that the larger masses near the encampment were broken before the morning of the 10th, but there was no change in the main body.

The Esquimaux revisited us in the morning, with their women and children; the party consisted of forty-eight persons. They seated themselves as before, in a semicircle, the men being in front, and the women behind. Presents were made to those who had not before received any; and we afterwards purchased several pairs of seal-skin boots, a few pieces of dressed seal-skin, and some deer-skin cut and twisted, to be used as cords. Beads, pins, needles, and ornamental articles, were most in request by the women, to whom the goods principally belonged, but the men were eager to get any thing that was made of iron. They were supplied with hatchets, files, ice chisels, fire-steels, Indian awls, and fish-hooks. They were very anxious to procure knives, but as each was in possession of one, I reserved the few which we had for another occasion. The quarter from whence these knives were obtained, will appear in a subsequent part of the narrative. It was amusing to see the purposes to which they applied the different articles given to them; some of the men danced about with a large cod-fish hook dangling from the nose, others stuck an awl through the same part, and the women immediately decorated their dresses with the ear-rings, thimbles, or whatever trinkets they received. There was in the party a great proportion of elderly persons, who appeared in excellent health, and were very active. The men were stout and robust, and taller than Augustus, or than those seen on the east coast by Captain Parry. Their cheek-bones were less projecting than the representations given of the Esquimaux on the eastern coast, but they had the small eye, and broad nose, which ever distinguish that people. Except the young persons, the whole party were afflicted with sore eyes, arising from exposure to the glare of ice and snow, and two of the old men were nearly blind. They wore the hair on the upper lip and chin; the latter, as well as that on their head, being permitted to grow long, though in some cases a circular spot on the crown of the head was cut bare, like the tonsure of the Roman catholic clergy. Every man had pieces of bone or shells thrust through the septum of his nose; and holes were pierced on each side of the under lip, in which were placed circular pieces of ivory, with a large blue bead in the centre, similar to those represented in the drawings of the natives on the N.W. coast of America, in Kotzebue's Voyage. These ornaments were so much valued, that they declined selling them; and when not rich enough to procure beads or ivory, stones and pieces of bone were substituted. These perforations are made at the age of puberty; and one of the party, who appeared to be about fourteen years old, was pointed out, with delight, by his parents, as having to undergo the operation in the following year. He was a good-looking boy, and we could not fancy his countenance would be much improved by the insertion of the bones or stones, which have the effect of depressing the under lip, and keeping the mouth open.

Their dress consisted of a jacket of rein-deer skin, with a skirt behind and before, and a small hood; breeches of the same material, and boots of seal-skin. Their weapons for the chase were bows and arrows, very neatly made; the latter being headed with bone or iron; and for fishing, spears tipped with bone. They also catch fish with nets and lines. All were armed with knives, which they either keep in their hand, or thrust up the sleeve of their shirt. They had received from the Loucheux Indians some account of the destructive effects of guns. The dress of the women differed from that of the men only in their wearing wide trowsers, and in the size of their hoods, which do not fit close to the head, but are made large, for the purpose of receiving their children. These are ornamented with stripes of different coloured skins, and round the top is fastened a band of wolf's hair, made to stand erect. Their own black hair is very tastefully turned up from behind to the top of the head, and tied by strings of white and blue beads, or cords of white deer-skin. It is divided in front, so as to form on each side a thick tail, to which are appended strings of beads that reach to the waist. The women were from four feet and a half to four and three quarters high, and generally fat. Some of the younger females, and the children, were pretty. The lady whose portrait adorns this work, was mightily pleased at being selected by Lieutenant Back for his sketch, and testified her joy by smiles and many jumps. The men, when sitting for their portraits, were more sedate, though not less pleased, than the females; some of them remarked that they were not handsome enough to be taken to our country. It will be seen from the engraving, that one of the men had a different cast of countenance from the rest; we supposed him to be descended from the Indians.

It would appear that the walrus does not visit this part of the coast, as none of these people recognised a sketch of one, which Lieutenant Back drew; but they at once knew the seal and rein-deer. We learned that the polar bear is seldom seen, and only in the autumn; and likewise that there are very few of the brown bears, which we frequently saw on the coast eastward of the Copper-mine River. We had already seen a few white whales, and we understood that they would resort to this part of the coast in greater numbers with the following moon.

The habits of these people were similar, in every respect, to those of the tribes described by Captain Parry, and their dialect differed so little from that used by Augustus, that he had no difficulty in understanding them. He was, therefore, able to give them full particulars relative to the attack made by the other party, and they expressed themselves much hurt at their treacherous conduct. " Those are bad men," they said, " and never fail either to quarrel with us, or steal from us, when we meet. They come, every spring, from the eastern side of the Mackenzie, to fish at the place where you saw them, and return as soon as the ice opens. They are distinguished from us, who live to the westward of the river, by the men being tattoed across the face. Among our tribes the women only are tattoed ;" having five or six blue lines drawn perpendicularly from the under lip to the chin. The speaker added, " If you are obliged to return by this way, before these people remove, we, with a reinforcement of young men, will be in the vicinity, and will willingly accompany you to assist in repelling any attack." Augustus returned with the Esquimaux to their tents, as there was not the least prospect of our getting forward, though the ice was somewhat broken.

A strong breeze from the westward during the night, contributed, with the swell, to the further reduction of the ice, in front of the encampment; and on the morning of the 11th, the wind changed to the eastward, and removed the pieces a little way off shore, though they were tossing too violently for the boats to proceed. The swell having subsided in the afternoon, we embarked; but at the end of a mile and a half were forced to land again, from the ice being fixed to the shore; and as the wind had now become strong, and was driving the loose pieces on the land the boats were unloaded and landed on the beach. From the summit of an adjoining hill we perceived an unbroken field of ice to the west, and, consequently, a barrier to our progress.

We encamped on the spot which our Esquimaux friends had left in the morning, to remove in their oomiacks and kaiyacks towards the Mackenzie, where they could set their fishing nets, and catch whales and seals. One of them showed his honesty, by returning some arrows, and a piece of a pemmican bag, that we had left at our last resting-place. The men also joined us here with specimens of rock from Mount Fitton.

The Esquimaux winter residences at this spot were constructed of drift timber, with the roots of the trees upwards, and contained from one to three small apartments, beside a cellar for their stores. There were generally two entrances, north and south, so low as to make it necessary to crawl through them. The only other aperture was a hole at the top for the smoke, which, as well as the doorways, could be filled up with a block of snow at pleasure. When covered with snow, and with lamps or fire burning within, these habitations must be extremely warm, though to our ideas rather comfortless. Lofty stages were erected near them for the purpose of receiving their canoes, and bulky articles. The annexed engraving from Lieutenant Back's sketch of these habitations, renders further description unnecessary. A north-east gale came on in the evening, and rolled such a heavy surf on the beach, that twice, during the night, we were obliged to drag the boats and cargoes higher up.

About three the next morning a heavy rain commenced, and continued, without intermission, through the day; at which we were delighted, however comfortless it made our situation, because we saw the ice gradually loosening from the land under its effects. We found the keeping a tide-pole fixed in the loose gravel beach impracticable here, as well as at the last resting-place, on account of the swell. It appeared to be high water this morning at half past one A.M., and that the rise of tide was about two feet. I need hardly observe that we had the sun constantly above the horizon, were it not for the purpose of

R

mentioning the amusing mistakes which the men made as to the hour. In fact, when not employed, a question as to the time of day never failed to puzzle them, except about midnight, when the sun was near the northern horizon.

Lieutenant Back missing the protractor which he used for laying down his bearings on the map, Augustus set off in the rain early this forenoon to recover it from an Esquimaux woman, whom he had seen pick it up. The rain ceased in the afternoon, the wind gradually abated, and by eight in the evening it was calm. A south wind followed, which opened a passage for the boat, but Augustus was not in sight. At midnight we became greatly alarmed for his safety, having now found that he had taken his gun, which we supposed the natives might have endeavoured to wrest from him, and we were on the point of despatching a party in search of him, when he arrived at four in

Thursday, 13th.   the morning of the 13th, much fatigued, accompanied by three of the natives. His journey had been lengthened by the Esquimaux having gone farther to the eastward than he had expected, but he had recovered the protractor, which had been kept in their ignorance of its utility to us. His companions brought five white fish, and some specimens of crystal, with other stones, from the mountains, which we purchased, and further rewarded them for their kindness in not allowing Augustus to return alone.

The boats were immediately launched, and having pulled a short distance from the land, we set the sails, our course being directed to the outer point in view, to avoid the sinuosities of the coast. We passed a wide, though not deep bay, whose points were named after my friends Captains Sabine and P. P. King; and we were drawing near the next projection, when a compact body of ice was discovered, which was joined to the land ahead. At the same time a dense fog came on, that confined our view to a few yards; it was accompanied by a gale from the land, and heavy rain. We had still hopes of getting round the point, and approached the shore in that expectation, but found the ice so closely packed that we could neither advance nor effect a landing. We, therefore, pulled to seaward, and turned the boat's head to the eastward, to trace the outer border of the ice. In this situation we were exposed to great danger from the sudden change of wind to S.E., which raised a heavy swell, and brought down upon us masses of ice of a size that, tossed as they were by the waves, would have injured a ship. We could only catch occasional glimpses of the land through the fog, and were kept in the most anxious suspense, pulling in and out between the floating masses of ice, for five hours, before we could get near the shore. We landed a little to the west of Point Sabine, and only found sufficient space for the boats and tents between the bank and the water. The rain ceased for a short time in the evening, and, during this interval, we perceived, from the top of the bank, that the whole space between us and the distant point, as well as the channel by which we had advanced to the westward, were now completely blocked; so that we had good reason to congratulate ourselves on having reached the shore in safety.

R 2

CHAPTER IV.

Babbage River—Meet Natives at Herschel Island—Their Trade with the Russians, through the Western Esquimaux—Ascend Mount Conybeare—Boundary of the British Dominions on this Coast—Delayed at Icy Reef—Barter Island—Detention at Foggy Island—Return Reef—Limit of outward Voyage.

1826.
July.

Friday,
14th.

ALTHOUGH it rained heavily during the night, and the wind blew strong off the land for some hours, there was no other change in the state of the ice on the morning of the 14th, than that the smaller pieces were driven a short way from the beach. The day was foggy and rainy, but the evening fine. The bank under which we were encamped is of the same earthy kind as that described on the 9th, but rather higher and steeper. It contains much wood-coal, similar to that found in the Mackenzie River, and at Garry's Island. The beach and the beds of the rivulets that flow through the ravines, consist of coarse gravel. Specimens of its stones, of the coal, and of the plants in flower, were added to the collection. We saw two marmots, and two rein-deer, which were too wary to allow of our getting within shot of them. Between noon and ten P.M. the loose ice was driving in front of the encampment from the N.W. to S.E., and at the latter hour it stopped. We could not detect any difference in the height of the water, and there was a calm the whole time. A light breeze from S.E. after midnight, brought the masses close to the beach.

Saturday,
15th.

On the morning of the 15th, having perceived that the ice was loosened from the land near the outer point, to which I have given the name of Kay, after some much-esteemed relatives, we embarked, and in the course of a few hours succeeded in reaching it, by passing between the grounded masses of ice. On landing at Point Kay, we observed that our progress must again be stopped by a compact body of ice that was fast to the shore of a deep bay, and extended to our utmost view seaward; and that we could not advance farther than the mouth of a river which discharged its waters just round the point. The boats were, therefore, pulled to its entrance, and we encamped. Former checks had taught us to be patient, and we, therefore, commenced such employments as would best serve to beguile the time, consoling ourselves with the hope that a strong breeze would soon spring up from the land and open a passage. Astronomical observations were obtained, the map carried on, and Lieutenant Back sketched the beautiful scenery afforded by a view of the Rocky Mountains, while I was employed in collecting specimens of the plants in flower. The men amused themselves in various ways, and Augustus went to visit an Esquimaux family that were on an island contiguous to our encampment.

We now discovered that the Rocky Mountains do not form a continuous chain, but that they run in detached ranges at unequal distances from the coast. The Richardson chain commencing opposite the mouth of the Mackenzie, terminates within view of our present situation. Another range, which I have named in honour of Professor Buckland, begins on the western side of Phillips Bay, and extending to the boundary of our view, is terminated by the Conybeare Mountain.

It gave me great pleasure to affix the name of my friend Mr. Babbage to the river we had discovered, and that of Mr. Phillips, Professor of Painting at the Royal Academy, to the bay into which its waters are emptied. We learned from the Esquimaux that this river, which they call Cöok-Keaktok, or Rocky

River, descends from a very distant part of the interior, though they are unacquainted with its course beyond the mountains. It appeared to us to flow between the Cupola and Barn mountains of the Richardson chain. There are many banks of gravel near its mouth, but above these obstructions the channel appeared deep, and to be about two miles broad. There were no rocks *in situ*, or large stones, near the encampment; the rolled pebbles on the beach were sandstone of red and light-brown colours, greenstone, and slaty limestone. We gathered a fine specimen of tertiary pitch-coal.

Augustus returned in the evening with a young Esquimaux and his wife, the only residents at the house he had visited. They had now quite recovered the panic into which they had been thrown on our first appearance, which was heightened by their being unable to escape from us owing to the want of a canoe. We made them happy by purchasing the fish they brought, and giving them a few presents; they continued to skip and laugh as long as they staid. The man informed us that judging from the rapid decay of the ice in the few preceding days, we might soon expect it to break from the land, so as to allow of our reaching Herschel Island, which was in view; but he represented the coast to the westward of the island as being low, and so generally beset with ice, that he was of opinion we should have great difficulty in getting along. This couple had been left here to collect fish for the use of their companions, who were to rejoin them for the purpose of killing whales, as soon as the ice should break up; and they told us the black whales would soon come after its rupture took place. It would be interesting to ascertain where the whales retire in the winter, as they require to inhale the air frequently. Those of the white kind make their appearance when there are but small spaces of open water; and we afterwards saw two black whales in a similar situation. One might almost infer from these circumstances that they do not remove very far. Is it probable that they go, at the close of the autumn, to a warmer climate? or can the sea be less closely covered with ice in the high northern latitudes? The situation of our encampment was observed to be, latitude 69° 19′ N.; longitude 138° 10½′ W.; variation 46° 16′ E.; and a rise and fall of nine inches in the water. The wind blew from the west during the night, and drove much ice near the boats; but as the masses took ground a little way from the shore, we were spared the trouble of removing the boats higher up the beach.

We were favoured in the forenoon of the 16th by a strong breeze from the land, which, in the course of a few hours, drove away many of these pieces towards Point Kay, and opened a passage for the boats. We immediately embarked to sail over to the western side of Phillips Bay, concluding, from the motion of the ice, that it must now be detached from that shore. On reaching it, we had the pleasure of finding an open channel close to the beach, although the entrance was barred by a stream of ice lying aground on a reef. The boats being forced by poles over this obstruction, we stood under sail along the coast to about five miles beyond Point Stokes; but there we were again compelled by the closeness of the ice to stop, and from the top of a sand-hill we could not discover any water in the direction of our course. The tents were therefore pitched, and the boats unloaded, and hauled on the beach. Heavy rain came on in the evening, by which we indulged the hope that the ice might be loosened. We were encamped on a low bank of gravel which runs along the base of a chain of sand-hills about one hundred and fifty feet high, and forms the coast line. The bank was covered with drift timber, and is the site of a deserted Esquimaux village. The snow still remaining in the ravines was tinged with light red spots. The night was calm, and the ice remained in the same fixed state until six in the morning of the 17th, when, perceiving the pieces in the offing to be in motion we launched the boats, and by breaking our way at first with hatchets, and then forcing with the poles through other streams of ice, we contrived to reach some lanes of water, along which we navigated for four hours. A strong breeze springing up from seaward, caused the ice to close so fast upon the boat, that we were obliged to put again to the shore, and land on a low bank, similar to that on which we had rested the night before. It was intersected, however, by many pools and channels of water, which cut off our communication with the land. As we could not obtain, from our present station, any satisfactory view of the state of the ice to the westward, I despatched Duncan and Augustus to take a survey of it from Point Catton, while Lieutenant Back and I made some astronomical observations. They returned after an absence of two hours, and reported that there was water near Herschel Island, and a channel in the offing that appeared to lead to it. We, therefore, embarked; and by pushing the boats between the masses that lay aground, for some distance, we succeeded in reaching open water at the entrance of the strait which lies between the island and the main, and through which the loose pieces of ice were driving fast to the westward. Having now the benefit of a strong favourable breeze, we were enabled to keep clear of them, and made good progress. Arriving opposite the S.E. end of Herschel Island, we perceived a large herd of rein-deer just taking the water, and on approaching the shore to get within shot, discovered three Esquimaux in pursuit. These men stood gazing at the boats for some minutes, and after a short consultation, we observed them to change the heads of their arrows, and prepare their bows. They then walked along the south shore, parallel to our course, for the purpose, as we soon found, of rejoining their wives. We reached the place at which the ladies were before them, and though invited to land, we were not able, on account of the surf. Augustus was desired to assure them of our friendship, and of our intention to stop at the first sheltered spot, to which they and their husbands might come to receive a present. More than this our little friend could not be prevailed upon to communicate, because they were " old wives;" and it was evident that he considered any further conversation with women to be beneath his dignity. On passing round the point we discovered that the ice was closely packed to leeward, and such a heavy swell setting upon it, that it was unsafe to proceed. We, therefore, encamped, and Augustus set off immediately to introduce himself to the Esquimaux. The tents were scarcely pitched, and the sentinels placed, before he returned, accompanied by twelve men and women, each bringing a piece of dried meat, or fish, to present to us. We learned from them that the boats, when at a distance, had been taken for pieces of ice; but when we drew near enough for them to distinguish the crews, and they perceived them clothed differently from any men they had seen, they became alarmed, and made ready their arrows, as we had observed. On receiving some presents, they raised a loud halloo, which brought five or six others from an adjoining island, and in the evening there was a further addition to the party of some young men, who had been hunting, and who afterwards sent their wives to bring us a part of the spoils of their chase. They remained near the tents the greater part of the night, and testified their delight by dancing and singing. An old woman, whose hair was silvered by age, made a prominent figure in these exhibitions.

The information we obtained from them confirmed that which

S

we had received from the last party, namely, that they procure the iron, knives, and beads, through two channels, but principally from a party of Esquimaux who reside a great distance to the westward, and to meet whom they send their young men every spring with furs, seal-skins, and oil, to exchange for those articles; and also from the Indians, who come every year from the interior to trade with them by a river that was directly opposite our encampment; which I have, therefore, named the Mountain Indian River. These Indians leave their families and canoes at two days' march from the mouth of the river, and the men come alone, bringing no more goods than they intend to barter. They were represented to be tall stout men, clothed in deer-skins, and speaking a language very dissimilar to their own. They also said that the Esquimaux to the westward, speak a dialect so different from theirs, that at the first opening of the communication, which was so recent as to be within the memory of two of our present companions, they had great difficulty in understanding them. Several quarrels took place at their first meetings, in consequence of the western party attempting to steal; but latterly there has been a good understanding between them, and the exchanges have been fairly made.

Our visitors did not know from what people either the Indians or the Esquimaux obtained the goods, but they supposed from some " Kabloonacht," (white people,) who reside far to the west. As the articles we saw were not of British manufacture, and were very unlike those sold by the Hudson's Bay Company to the Indians, it cannot be doubted that they are furnished by the Russian Fur Traders, who receive in return for them all the furs collected on this northern coast. Part of the Russian iron-work is conveyed to the Esquimaux dwelling on the coast east of the Mackenzie. The western Esquimaux use tobacco, and some of our visitors had smoked it, but thought the flavour very disagreeable. Until I was aware of their being acquainted with the use of it, I prohibited my men from smoking in their presence, and afterwards from offering their pipes to the Esquimaux at any time. At the conclusion of this conference, our visitors assured us, that having now become acquainted with white people, and being conscious that the trade with them would be beneficial, they would gladly encourage a further intercourse, and do all in their power to prevent future visitors from having such a reception as we had on our arrival in these seas. We learned that this island, which has been distinguished by the name of Herschel, is much frequented by the natives at this season of the year, as it abounds with deer, and its surrounding waters afford plenty of fish. It is composed of black earth, rises, in its highest point, to about one hundred feet, and at the time of our visit was covered with verdure. The strait between it and the main shore, is the only place that we had seen, since quitting the Mackenzie, in which a ship could find shelter; but even this channel is much interrupted by shoals. Latitude 69° 33½′ N.; longitude 139° 3′ W.; were observed at the encampment.

On the morning of the 18th the fog was so thick that we could not see beyond the beach. It dispersed about noon, and we discovered that there was a channel of open water near the main shore, though in the centre of the strait the ice was heavy, and driving rapidly to the north-west. We embarked at once, in the expectation of being able to penetrate between the drift ice and the land, but the attempt was frustrated by the shallowness of the water; and the fog again spreading as thick as before, we landed on a sand-bank. We were soon visited by another party of the Esquimaux, who brought deer's meat for sale; and although the whole quantity did not amount to a deer, we had to purchase it in small pieces. This practice of dividing the meat among the party, we found to prevail throughout the

S 2

voyage; and they avowed as their reason for it, the desire that every one might obtain a share of the good things we distributed. One of the men drew on the sand a sketch of the coast to the westward, as far as he was acquainted with it; from which it appeared that there was a line of reefs in front of the coast the whole way; the water being deep on the outside of them, but on the inside too shallow even for their oomiacks to float. We subsequently found that his knowledge of the coast did not extend beyond a few days' march.

The atmosphere becoming more clear about two P.M., we again embarked, and endeavoured to get to seaward. The boats, however, soon grounded; and finding all our attempts to push through any of the channels between the reefs ineffectual, we pulled back close to Herschel Island. Following, then, the course of the drift ice, we passed near to its south-west point, which was found to be the only deep passage through the strait. We afterwards entered into a fine sheet of open water, the main body of the ice being about half a mile to seaward, and only a few bergs lying aground in the direction of our course. The outer parts of the island appeared closely beset with ice. At the end of five miles we discerned another large party of Esquimaux, encamped on a reef; they waved their jackets as signals for us to land, which we declined doing, as we perceived the water to be shallow between us and them. They ran along the beach as far as the end of the reef, tempting us by holding up meat. Only two of the party were provided with canoes, and they followed us to a bluff point of the main shore, on which we landed. These proved to be persons whom we had seen at Herschel Island, and who had visited the Esquimaux in this quarter on purpose to make them acquainted with our arrival. We were happy to learn from them that we should not see any more of their countrymen for some time, because, while

1826.
July.

surrounded by them, the necessity of closely watching their motions, prevented us from paying due attention to other objects. Resuming our voyage, we pulled along the outer border of a gravel reef, about two hundred yards broad, that runs parallel to, and about half a mile from, the coast, having a line of drift ice on the outside of us. The wind being contrary, and the evening cold, temperature 40°, we encamped on the reef at eight P.M., where we found plenty of drift timber; the water was brackish. The distance travelled this day was eight miles and a half. The main shore opposite the encampment was low to a great distance from the coast; it then appeared to ascend gradually to the base of the Buckland chain of mountains.

The following morning being calm, and very fine, the boats were launched at three A.M., and we set off in high spirits; but after pulling three miles, we perceived the channel of open water becoming narrow, and the pieces of ice heavier than any we had before seen, some of them being aground in three fathoms water. At six A.M., after having gone five miles and a half, we were stopped by the ice which adhered to the reef, and was unbroken to seaward. Imagining we saw water at some distance beyond this barrier, we were induced to drag the boats across the reef, and launch them into the channel on the inside, in the hope of reaching it. This proved to be a bay, at the head of which we arrived in a short time. It was then discovered that a fog hanging over the ice had been mistaken for water. The boats were, therefore, reconveyed across the reef, the tents pitched, and we had to draw largely on our nearly exhausted stock of patience, as we contemplated the dreary view of this compact icy field. A herd of rein-deer appeared very opportunely to afford some employment, and most of the men were despatched on the chase, but only one was successful. The following observations were obtained:—Latitude 69° 36′ N.; longitude 139° 42′ W.; variation

Wednesday, 19th.

46° 13′ E. Being now abreast of Mount Conybeare, Lieutenant Back and I were on the point of setting out to visit its summit, when we were stopped by a very dense fog that accompanied a fresh breeze from the N.W., followed by heavy rain. The weather continued bad, until ten the following morning; the ice near the beach was broken into smaller pieces, but as yet too closely packed for our proceeding. The water being brackish in front of the reef, we despatched two men to bring some from the pools at a distance inland, which was found to have the same taste; from this circumstance, as well as from the piles of drift wood, thrown up far from the coast, one may infer that the sea occasionally washes over this low shore. The ice broken off from large masses, and permitted to drain before it was melted, did not furnish us with better water. A couple of pin-tailed ducks were shot, the only pair seen; the black kind were more numerous, but were not fired at, as they are fishing ducks, and, therefore, not good to eat. We also saw a few geese and swans.

The atmosphere was calm, and perfectly clear, on the morning of the 21st; and as there was not any change in the position of the ice, I visited Mount Conybeare, accompanied by Duncan and Stewart. Though its distance was not more than twelve miles from the coast, the journey proved to be very fatiguing, owing to the swampiness of the ground between the mountain and the sea. We had also the discomfort of being tormented the whole way by myriads of musquitoes. The plain was intersected by a winding river, about forty yards broad, which we forded, and on its western side found a thicket of willows, none of which were above seven inches in circumference, and only five or six feet high. At the foot of the mountain were three parallel platforms, or terraces, whose heights we estimated at fifty, eighty, and one hundred and thirty feet; composed of transition slate, the stone of the lowest being of the closest texture. We found the task of climbing above the upper terraces difficult, in consequence of the looseness of the stones, which did not afford a firm footing, but after an hour's labour, we succeeded in reaching the top. The mountain is also composed of slate, but so much weathered near the summit, as to appear a mere collection of stones. Its height above the sea we estimated at eight hundred feet. Two or three hardy plants were in flower, at the highest elevation, which we gathered, though they were of the same kind that had been collected in the lower lands; and during the whole march we did not meet with any plant different from the specimens we had already obtained. On arriving at the top of the mountain, we were refreshed by a strong south wind, which we fondly hoped might reach to the coast, and be of service, by driving the ice from the land. This hope, however, lasted only a few minutes; for, on casting our eyes to seaward, there appeared no open water into which it could be moved, except near Herschel Island. The view into the interior possessed the charm of novelty, and attracted particular regard. We commanded a prospect over three ranges of mountains, lying parallel to the Buckland chain, but of less altitude. The view was bounded by a fourth range of high-peaked mountains, for the most part covered with snow. This distant range was afterwards distinguished by the name of the British Chain; and the mountains at its extremities were named in honour of the then Chancellor of the Exchequer, and President of the Board of Trade—the Right Honourable Mr. Robinson, now Lord Goderich, and Mr. Huskisson. When seen from the coast, the mountains of the Buckland chain appeared to form a continuous line, extending from N.W. by N., to S.E. by S.; but from our present situation we discovered that they were separated from each other by a deep valley, and a rivulet, and that their longest direction was N.N.E. and S.S.W. The same order prevailed in the three ranges behind the Buckland chain; and the highest of their mountains, like Mount Conybeare, were round and naked at the top; the vallies between them were grassy. We erected a pile of stones of sufficient height to be seen from the sea, and deposited underneath it a note, containing the latitude, longitude, and some particulars relative to the Expedition.

The 22nd was a calm sultry day, the temperature varying between 58° and 63°, and we were tormented by musquitoes. The ice remained very close to the beach. Impatient of our long detention, we gladly availed ourselves, at three in the morning

of the 23rd, of a small opening in the ice, to launch the boats, and push them forward as far as we could get them. We thus succeeded in reaching a lane of water, through which we made tolerable progress, though after two hours and a half of exertion, we were gradually hemmed in, and forced again to encamp at the mouth of a small stream westward of Sir Pulteney Malcolm River. We had, however, the satisfaction of finding, by the observations, that we had gained ten miles. Latitude 69° 36′ N.; longitude 140° 12′ W.; variation 45° 6′ E. The temperature of the water at the surface a quarter of a mile from the shore was 40°, that of the air being 49°. The water was two fathoms deep, ten yards from the beach.

The coast here was about fifteen feet high; and from the top of the bank a level plain extended to the base of the mountains, which, though very swampy, was covered with verdure. At this place we first found boulder stones, which were deeply seated in the gravel of the beach. They consisted of greenstone, sandstone, and limestone; the first mentioned being the largest, and the last the most numerous. Having seen several fish leaping in the river, a net was set across its mouth, though without success, owing to the meshes being too large. Two men were despatched to examine the state of the ice; and on their return from a walk of several miles, they reported that, with the exception of a small spot close to the beach, it was quite compact. They had observed, about two miles from the encampment, stumps of drift wood fixed in the ground at certain distances, extending from the coast across the plain towards the Rocky Mountains, in the direction of two piles of stones, which were erected on the top of the latter. We were at a loss to conjecture what motive the Esquimaux could have had for taking so much trouble, unless these posts were intended to serve as decoys for the rein-deer. The party assembled at divine service in the evening, as had been our practice every Sunday.

On the morning of the 24th we were able to make a further advance of two miles and three quarters, by forcing the boats between the masses of ice, as far as the debouche of another rivulet, in latitude 69° 36½′ N., and longitude 140° 19½′ W. Under any other circumstance than that of being beset by ice, the beautifully calm and clear weather we then had would have been delightful; but as our hope of being released rested solely on a strong wind, we never ceased to long for its occurrence. A breeze would have been, at any rate, beneficial in driving away the musquitoes, which were so numerous as to prevent any enjoyment of the open air, and to keep us confined to a tent filled with smoke, the only remedy against their annoyance.

We were still detained the two following days, and the only things we saw were a grey wolf, some seals, and some ducks. More tedious hours than those passed by us in the present situation, cannot well be imagined. After the astronomical observations had been obtained and worked, the survey brought up, a sketch made of the encampment, and specimens of the plants and stones in the vicinity collected, there was, literally, nothing to do. The anxiety which was inseparable from such an enterprise as ours, at such an advanced period of the season, left but little disposition to read, even if there had been a greater choice of books in our travelling library, and still less composure to invent amusement. Even had the musquitoes been less tormenting, the swampiness of the ground, in which we sank ancle deep at every step, deprived us of the pleasure of walking. A visit to the Rocky Mountains was often talked of, but they were now at a distance of two days' journey, and we dared not to be absent from the boats so long, lest the ice, in its fickle movements, should open for a short time. Notwithstanding the closeness of the ice, we perceived a regular rise and fall of the water, though it amounted only to seven inches, except on the night of the 24th, when the rise was two feet; but the direction of the flood was not yet ascertained. We found a greater proportion of birch-wood, mixed with the drift timber to the westward of the Babbage, than we had done before; between the Mackenzie and that river it had been so scarce, that we had to draw upon our store of bark to light the fires. Some lunar observations were obtained in the afternoon of the 25th, and their results assured us that the chronometers were going steadily. At midnight we were visited by a strong S.W. breeze, accompanied by rain, thunder, and lightning. This weather was succeeded by calm, and a fog that continued throughout the next day, and confined our view to a few yards. Temperature from 41° to 43°. On the atmosphere becoming clear about nine

in the evening of the 26th, we discovered a lane of water, and immediately embarking, we pulled, for an hour, without experiencing much interruption from the ice. A fresh breeze then sprung up from the N.W., which brought with it a very dense fog, and likewise caused the ice to close so fast upon us, that we were compelled to hasten to the shore. We had just landed, when the channel was completely closed. We encamped on the western side of a river about two hundred yards broad, which, at the request of Lieutenant Back, was named after Mr. Backhouse, one of the under Secretaries of State for Foreign Affairs. It appeared that the water that flowed from this channel had caused the opening by which we had travelled from our last resting-place; for beyond it, the ice was closely packed.

Some heavy rain fell in the night, and the morning of the 27th was foggy; but the sun, about noon, having dispersed the fog, we discovered an open channel, about half a mile from the shore. No time was lost in pushing the boats into it. By following its course to the end, and breaking our way through some streams of ice, we were brought, at the end of eight miles, to the mouth of a wide river that flows from the British range of mountains. This being the most westerly river in the British dominions on this coast, and near the line of demarcation between Great Britain and Russia, I named it the Clarence, in honour of His Royal Highness the Lord High Admiral. Under a pile of drift timber which we erected on the most elevated point of the coast near its mouth, was deposited a tin box, containing a royal silver medal, with an account of the proceedings of the Expedition; and the union flag was hoisted under three hearty cheers, the only salute that we could afford. This ceremony did not detain us longer than half an hour; when we launched into a larger space of open water than we had seen since the 9th of the month. This circumstance, together with the appearance of several seals, and the water becoming more salt, created a hope that we should soon enter upon a brisker navigation. But this too sanguine expectation was dispelled in little more than an hour, by a close and heavy field of ice, which obliged us to pull to the shore. The tent was pitched under a steep bank of mud, in latitude 69° 38′ N.; longitude 140° 46′ W. The soundings this day

T 2

varied from two to ten fathoms ; and the temperature of the air from 37° to 45°. The ice having opened near the beach by noon of the 28th, so as to admit the boats, we embarked, to try if we could not advance by thrusting the masses aside with poles. After spending several hours in this labour, and gaining only two miles, further exertion became ineffectual, owing to the ice being closely packed, and many of the pieces from fifteen to twenty feet high, lying aground. We had, however, gained by the removal the comforts of dry ground, and good water, which had been wanting at the last encampment. Among the drift timber. on the beach was a pine tree, seven feet and a quarter in girth, by thirty-six long. We had previously seen several, little inferior in size. The temperature this day varied from 39° to 48°. We had observed, for the preceding fortnight, that the musquitoes assailed us as soon as the temperature rose to 45°, and that they retired quickly on its descending below that height.

The morning of the 29th opened with heavy rain and fog; the precursors of a strong gale from E.N.E., which brought back the ice we had already passed, and closely packed it along the beach, but we could not perceive that the wind had the slightest effect on the main body at a distance from the shore. This was a very cold, comfortless day, the temperature between 38° and 42°.

On the following morning a brilliant sun contributed with the gale to the dispersion of the mist which had, for some days past, overhung the Rocky Mountains, and we had the gratification of seeing, for the first time, the whole length of the British Chain of Mountains, which are more peaked and irregular in their outline, and more picturesque than those of the Buckland Range. The following observations were obtained here :—Latitude 69° 38′ N.; longitude 140° 51′ W.; variation 45° 43′ E.; Dip 83° 27′. In exploring the bed of a rivulet we found several pieces of quartz, containing pyrites of a very bright colour, which so much attracted the attention of the crews, that they spent several hours in examining every stone, expecting to have their labour rewarded by the discovery of some precious metal.

The gale having abated in the evening, we quickly loaded the boats, and pulled them into a lane of water that we had observed about half a mile from the shore. This, however, extended only a short way to the west, and at the end of a mile and a half inclined towards the beach, the ice beyond it being closely packed. Before the boats could be brought to the land, they received several heavy blows in passing through narrow channels, and over tongues of grounded ice. I walked to the extreme point that we had in view from the tent, and was rejoiced by the sight of a large space of water in the direction of our course; but up to the point the ice was still compact, and heavy. On my way I passed another Esquimaux village, where there were marks of recent visitors.

We witnessed the setting of the sun at eleven P.M.; an unwelcome sight, which the gloomy weather had, till then, spared us; for it forced upon our minds the conviction that the favourable season for our operations was fast passing away, though we had, as yet, made so little progress. This was not the only uncomfortable circumstance that attended us this evening. Our friend Augustus was seized with a shivering fit, in consequence of having imprudently rushed, when in full perspiration, into a lake of cold water, to drag out a rein-deer which he had killed. He was unable to walk on coming out of the water, and the consequence would have been more serious had it not been for the kindness of his companion, Wilson, who deprived himself of his flannels and waistcoat to clothe him. On their arrival at the tent, Augustus was put between blankets, and provided with warm chocolate, and the only inconvenience that he felt next morning was pain in his limbs.

We had several showers of rain during the night, with a steady S.W. breeze, and in the morning of the 31st were delighted by perceiving the ice loosening and driving off the land. We were afloat in a few minutes, and enjoyed the novelty of pulling through an uninterrupted channel as far as Point Demarcation, which has been so named from its being situated in longitude 141° W., the boundary between the British and Russian dominions on the northern coast of America. This point seems to be much resorted to by the Esquimaux, as we found here many winter houses, and four large stages. On the latter were deposited several bundles of seal and deer skins, and several pair of snow-shoes. The snow-shoes were netted with cords of deer-skin, and were shaped like those used by the Indians near the Mackenzie. A favourable breeze now sprang up; and having ascertained, by mounting one of the Esquimaux stages, that there was still a channel of open water between a low island and the main shore, we set sail to follow its course. At the end of three miles we found the water gradually to decrease from three fathoms to as many feet, and shortly afterwards the boats repeatedly took the ground. In this situation we were enveloped by a thick fog, which limited our view to a few yards. We, therefore, dragged the boats to the land, until we could see our way; this did not happen before ten in the evening, when it was discovered from the summit of an eminence about two miles distant, that though the channel was of some extent, it was very shallow, and seemed to be barred by ice to the westward. We also ascertained that it was bounded to seaward by a long reef. The night proved very stormy, and we were but scantily supplied with drift wood.

Though the morning of the 1st of August commenced with a heavy gale from E.N.E., and very foggy weather, we proceeded to the reef, after much fatigue in dragging the boats over the flats, under the supposition that our best chance of getting forward would be by passing on the outside of it. But there finding heavy ice lying aground, and so closely packed as to preclude the possibility of putting the boats into the water, it was determined to examine the channel by walking along the shore of the reef. An outlet to the sea was discovered, but the channel was so flat that gulls were, in most parts, wading across it; and there was, therefore, no other course than to await the separation of the ice from the reef. On the dispersion of the fog in the afternoon, we perceived that some of the masses of ice were from twenty to thirty feet high; and we derived little comfort from beholding, from the top of one of them, an unbroken surface of ice to seaward.

The gale blew without the least abatement throughout the night, and until noon of the 2nd, when it terminated in a violent gust, which overthrew the tents. The field of ice was broken in the offing, and the pieces put in motion; and in the evening there appeared a large space of open water, but we could not take advantage of these favourable circumstances, in consequence of the ice still closely besetting the reef. Lieutenant Back occupied himself in sketching the different views from the reef; from one of which the annexed engraving has been selected, as conveying an accurate delineation of our position on Icy Reef. We remarked large heaps of gravel, fifteen feet above the surface of the reef, on the largest iceberg, which must have been caused by the pressure of the ice; and from the top of this berg we had the satisfaction of discovering that a large herd of reindeer were marching in line towards the opposite side of the channel. Our party was instantly on the alert, and the best hunters were sent in the Reliance in chase of them. The boat grounded about midway across, and the eager sportsmen jumped overboard and hastened to the shore; but such was their want of skill, that only three fawns were killed, out of a herd of three or four hundred. The supply, however, was sufficient for our present use, and the circumstances of the chase afforded amusing conversation for the evening. The astronomical observations place our encampment in latitude 69° 43′ N.; longitude 141° 30′ W. The temperature this day varied from 40° to 42°.

Thursday,
3rd.
On the morning of the 3rd a strong breeze set in from the east, which we were rejoiced to find caused a higher flood in the channel than we had yet seen, and the hope of effecting a passage by its course was revived; as the ice was still fast to the reef, and likely to continue so, it was considered better to occupy ourselves in dragging the boats through the mud, than to continue longer in this irksome spot, where the wood was already scarce, and the water indifferent. The boats, therefore, proceeded with four men in each, while the rest of the crew walked along the shore, and rendered assistance wherever it was necessary, to drag them over the shallow parts. After four hours' labour, we reached the eastern part of the bay, which I have had the pleasure of naming after my friend Captain Beaufort, R.N., and which was then covered with ice. We had also the happiness of finding a channel that led to seaward, which enabled us to get on the outside of the reef; but as we pushed as close as we could to the border of the packed ice, our situation, for the next four hours, was attended with no little anxiety. The appearance of the clouds bespoke the return of fog, and we were sailing with a strong breeze through narrow channels, between heavy pieces of drift ice, on the outside of a chain of reefs that stretched across Beaufort Bay, which we knew could not be approached within a mile, owing to the shallowness of the water.

Beyond Point Humphrys, the water being deep close to the coast, we travelled in more security, though the ice was less open than before. We halted to sup on a gravel reef that extends from the main shore to Point Griffin; having run twenty-eight miles, the greatest distance we had made on one day since our departure from the Mackenzie.

A black whale, and several seals, having been seen just before we landed, the water now decidedly salt, and the ice driving with great rapidity to the westward, were circumstances that we hailed with heartfelt joy; as affording the prospect of getting speedily forward, and in the evening we lost sight of Mount Conybeare, which had been visible since the 9th of July. There were several huts on the reef, and one large tent, capable of holding forty persons, which appeared to have been lately occupied, besides eighteen sledges, that we supposed to have been left by the men who had gone from Herschel Island, to exchange their furs with the western Esquimaux. Among the baggage we found a spoon, made out of the musk ox horn, like those used by the Canadian voyagers. At six this evening we passed the termination of the British chain of Mountains, and had now arrived opposite the commencement of another range, which I named after the late Count Romanzoff, Chancellor of the Russian Empire, as a tribute of respect to the memory of that distinguished patron and promoter of discovery and science.

Having taken the precaution of supplying ourselves with fresh water, we quitted the reef, to proceed on our voyage under sail, but shortly afterwards arrived at very heavy ice, apparently packed. We found, however, a narrow passage, and by forcing the boats through it, reached a more open channel, where the oars could be used. This extended along a reef, so that we could pursue our course with safety, being ready to land in the event of the ice drifting upon us. The sun set this evening at half past ten P.M.; and the temperature of the air during its disappearance was 38°. Between the reefs and the low main land the water was entirely free from ice. After passing Point Sir Henry Martin, we were tempted, by the appearance of a bay, to steer within the reefs, as we could then use the sails, and make a more direct course than by winding among the ice. The water proved so shallow that the boats took the ground, at the distance of three miles from the shore, which caused us to alter our plan, and follow the line of drift ice near the border of the pack. The breeze died away; and in proceeding under oars beyond Point Manning, we descried a collection of tents planted on a low island, with many oomiacks, kaiyacks, and dogs around them. The Esquimaux being fast asleep, Augustus was desired to hail them, and after two or three loud calls, a female appeared in a state of nudity; after a few seconds she called out to her husband, who awoke at the first sound of her voice, and shouting out that strangers were close at hand, the whole space between the tents and the water was, in a few minutes, covered with armed, though naked, people. Their consternation on being thus suddenly roused by strangers, of whose existence they had never heard, can be better imagined than described. We drew near the shore, to let Augustus inform them who we were, and of the purpose of our visit, which produced a burst of acclamation, and an immediate invitation to land. This we declined doing, having counted fifty-four grown persons, and knowing that we had not the means of furnishing such a number with the articles they might crave. Besides, it was evident, from their hurried manner, that they were in a state of high excitement, and might then, perhaps, have been disposed to seize upon every thing within their grasp. Four of the kaiyacks being launched, after we had receded to a proper distance from the island, we allowed them to come alongside; and presents were given to the men. We then learned that these were the people who had conveyed the furs, &c., from Herschel Island, and that the exchange with the western Esquimaux had been made at the place where they were encamped, only a few days before. They intended to commence their return this day to Herschel Island, where the iron and beads would be distributed among their relations, according to the furs, &c., they had supplied. The Esquimaux saluted us at parting with many vociferations of *teyma*, and we continued our journey for five miles; at the end of which, the wind setting in strongly against us, we landed at the western part of Barter Island, to refresh the crew. We then found that a rapid tide was running to the eastward, and at eleven the water had risen one foot, from the time of our landing. The tents were scarcely pitched before we saw two kaiyacks coming towards us from the westward, and the man in the headmost accepted, without hesitation, our invitation to land. His companion was asleep, and his canoe was driving with the wind and tide; but when awaked by the voice of Augustus, he also came. These were young men returning from hunting to the tents that we had passed; and being much fatigued, they made but a short stay. The only information collected from them was, that the coast before us was similar to that along which we had been travelling, and that the ice was broken from the shore. The latitude 70° 5′ N.; longitude 143° 55′ W.; variation 45° 36′ E.; were observed.

As soon as the latitude had been obtained, we embarked, favoured by wind and tide, to cross the bay, which has been named in honour of the Marquess Camden. The water was of a seagreen colour, perfectly salt, and from three to five fathoms deep; the temperature 35° at the surface, that of the air 43°. The day was very clear, and exposed to our view the outline of the Romanzoff chain of Mountains, whose lofty peaks were covered with snow. At the end of ten miles we observed four tents planted on a reef, and several women standing about them, who made many signs for us to land, but the surf was rolling too heavily on the beach. As we proceeded, their husbands were perceived on the main shore, in pursuit of a large herd of reindeer, which they seemed to be surrounding so as to drive the deer into the water, where they would probably spear them to more advantage.

Continuing along the shore beyond the reef at the distance of two miles from the land, the boats touched the ground several times, which made us conclude we were steering into a bay, though its outline could not be seen. The wind changed at the time to the north, blew strong, and raised a heavy swell, which induced us to haul out to seaward, and we soon afterward discovered an island, which we just reached under sail. From its summit we perceived a chain of low reefs, extending from its northern point for several miles to the westward, on which the wind was then blowing, and bringing down the drift ice. We were, therefore, compelled to halt, and await more moderate weather. This island, like the projecting points of the main shore, is a mere deposit of earthy mud, covered with verdure, about twenty or twenty-five feet high. There was another island adjoining, which seemed to be a collection of boulder stones ; from whence its designation on the map.

The ice appeared closely packed to seaward ; nearer to the island were icebergs aground, and within these, streams of loose pieces driving towards the reefs. In the hurry of embarkation from Barter Island, one of the crew of the Reliance left his gun and ammunition, which we regretted the more, from being apprehensive that an accident might happen to the natives. The circumstance was not known before the boats were a great distance from the island, or we should have put back to have recovered it.

A very thick fog came on in the evening. This weather, however, did not prevent our receiving a visit from two of the natives about midnight, who told Augustus that, having scented the smoke of a fire from the opposite side of the bay, they had come to ascertain who had made it. They were armed with bows and arrows, and advanced towards the tent without any alarm. We found that they had been hunting, with several other men, at the foot of the Romanzoff Mountains, and that they were now going to rejoin their friends at Barter Island, with the fruits of a successful chase. Their knowledge of the coast terminated at this place, which is as far to the westward as any of the party from Herschel Island travel.

The western Esquimaux had parted from them seven nights before, but they supposed that they had not made much progress, as their oomiacks were heavily laden. Those people had informed them that the coast to the westward was low, and fronted by reefs, like that we had already passed; the water also was very shallow; they therefore recommended that we should keep on the outside of every reef. Our visiters had no sooner received their presents than they raised a loud cry, which was intended to bring their friends. On the dispersion of the fog at the time, we discovered an oomiack, filled with people paddling, and some other men wading towards us. It being calm, and the swell having abated, we did not wait for their arrival, but embarked at one in the morning of the 5th, and pursued our course to the westward, keeping on the outside of the reefs. The water, however, was very shallow, even at the distance of two miles, and we were much teased by the boats repeatedly touching the ground. This was particularly the case when we arrived opposite to the large river, which was named in honour of the late Mr. Canning, where we found the water perfectly fresh, three miles from the land. The ice being more loose abreast of this river, we pulled out to seaward into deep water. The land was then hidden from our view by the haze, though not more than four miles distant, and our course was directed by the masses of ice lying aground; but at the end of three miles, our further progress was stopped at six A.M., by the ice being closely packed on the outer border of a reef, in latitude 70° 7′ N.; longitude 145° 27′ W.

We perceived, on landing, by the driving of the loose pieces of ice, that the tide was running strongly to the eastward, through the channel we had passed along, and that it continued to do so, until ten this morning, during which time the water was falling. It changed at ten, and the water rose one foot before one P.M. This observation would indicate the flood to come from the eastward, though contrary to what was remarked at Barter Island the day before; but in a sea so closely beset with ice, no accurate observations as to the direction of the tide could be obtained.

The Rocky Mountains either terminated abreast of our present situation, or receded so far to the southward as to be imperceptible from the coast a few miles beyond this reef. Lieutenant Back made a sketch of the most western mountain, which appears in the annexed vignette, and which I had the pleasure of naming in honour of Dr. Copleston, Provost of Oriel College, now Bishop of Landaff. The ice being somewhat loosened by the flood tide, we embarked at one P.M., to force the boats through the narrow channels, and in the course of two hours reached Point Brownlow, where we landed, for the purpose of ascertaining whether the ice could be avoided by passing into the bay that then opened to our view, trending to the south. We perceived that this bay was in every part flat, and strewed with stones; and that the only prospect of getting forward was by entering the ice again, and pushing to an island about two miles further to the west, which we reached after receiving several heavy blows in passing through the loose ice at the entrance of the strait, between the central reef and the island, where the pieces were much tossed by the tide.

The view from the south-east part of the island led us, at first, to suppose that we might proceed by keeping close to its south shore; but in making the attempt, the boats repeatedly took the ground, and we were obliged to seek a passage by the north side of the island. At the end of a mile in that direction we were stopped by the ice being unbroken from the shore, and closely packed to seaward, as is represented in the annexed engraving, from a sketch taken by Lieutenant Back, at the tents, about thirty feet above the horizon. Since the day after our departure from the Mackenzie, when we first came to the ice, we had not witnessed a more unfavourable prospect than that before us. No water was to be seen, either from the tents, or from the different points of the island which we visited, for the purpose of examining into the state of the ice. We were now scantily supplied with fuel; the drift timber being covered by the ice high up the bank, except just where the boat had landed.

In the evening a gale came on from the east, and blew throughout the following day: we vainly hoped this would produce some favourable change; and the water froze in the kettle on the night of the 5th. The position of the encampment was ascertained by observation to be, latitude 70° 11′ N.; longitude 145° 50′ W.; variation 42° 56′ E.; so that notwithstanding the obstructions we had met, an advance of two degrees of longitude had been made in the two preceding days.

This island received the name of Flaxman, in honour of the late eminent sculptor. It is about four miles long, and two broad, and rises, at its highest elevation, about fifty feet. In one of the ravines, where a portion of the bank had been carried away by the disruption of the ice, we perceived that the stratum of loose earth was not more than eighteen inches thick, the lower bed being frozen mud; yet this small quantity of soil, though very swampy, nourished grasses, several of the arctic plants, and some few willows, that were about three inches high. Several boulder stones were scattered on its beach, and also in the channel that separates it from the main shore.

An easterly wind gave place to a calm on the morning of the 7th: and as this change, though it produced no effect in loosening the ice to the north, caused more water to flow into the channel between the island and the main, we succeeded with little difficulty in crossing the flats that had before impeded us. Beyond this bar the water gradually deepened to three fathoms; and a favourable breeze springing up, we set the sail, and steered for the outer point of land in sight. We continued in smooth water until we reached Point Thomson, when, having lost the shelter of the ice which was aground on a tongue of gravel projecting from Flaxman Island, we became exposed to an unpleasant swell.

The Lion was very leaky, in consequence of the blows she had received from the ice; but as we could keep her free by baling, we did not lose the favourable moment by stopping to repair her. Our course was continued past Point Bullen, until we came to an island lying three miles from the shore, which proved to be connected with the main land by a reef. Dazzled by the glare of the sun in our eyes, the surf, which was breaking on this reef, was mistaken for a ripple of the tide; and although the sails were lowered, as a measure of precaution, we were so near before the mistake was discovered, that the strength of the wind drove the Lion aground, by which accident she took in much water. The exertions of the crew soon got her afloat, and both boats were pulled to windward of the island. The sails were then set, but as the wind had by this time increased to a strong gale, they were close reefed. We stood along the coast, looking for a favourable landing-place, that we might obtain shelter from an approaching storm which the appearance of the sky indicated, and to repair the damage which the Lion had sustained. At length, some posts that had been erected by the Esquimaux on a point, denoted an approachable part of the coast, and we effected a landing after lightening the boats, by carrying part of the cargo two hundred yards through the water. The main shore to the westward of Flaxman Island is so low that it is not visible at the distance of three miles, with the exception of three small hummocks, which look like islands.

The carpenter had finished the repairs of the boat by midnight, and we were prepared to go forward, but were prevented from moving by a very thick fog, which continued throughout the night, and till eleven on the morning of the 8th. The storm continued violent throughout the day, but the fog cleared away for the space of two hours, and enabled us to perceive that the ice, which in the preceding evening had been at a considerable distance from the land, was now tossing about, in large masses, close to the border of the shallow water. We were also enabled, during the interval of clear weather, to ascertain, by astronomical observations, the latitude 70° 16′ 27″ N.; longitude 147° 38′ W.; and variation 43° 15′ E.

The hunters were sent out in pursuit of some deer that were seen, and Augustus killed one. They ascertained, during the chase, that we were on an island, separated from the main shore by a channel, fordable at low water. At this encampment we remarked the first instance of regularity in the tide. It was low water at half past nine on the evening of the 7th, and high water at half past two the following morning; the rise being sixteen inches. An equally regular tide was observed on the 8th, but we could not ascertain the direction of the flood.

X

1826.
August.
Wednesday,
9th.

After sunset the squalls became extremely violent; and until three in the afternoon of the 9th, the fog was so dense that every object more distant than forty yards, was hidden. After that period, a partial clearness of the atmosphere discovered to us the waves more high than the day before, and beating heavily against the weather beach of the island. We rejoiced, however, at seeing a large stream of ice to windward, supposing that its presence there would cause the swell to go down, and that we should be able to proceed as soon as the wind should fall. We employed ourselves in observing the dip with Meyer's and the common needles, as well as the magnetic force. The mean dip was 82° 26'. The temperature of the air this day varied from 38° to 45°. High water took place at fifteen minutes after three P.M., the rise being two feet. The water did not fall so low as yesterday, owing to the wind blowing more across the mouth of the bay.

Thursday,
10th.

On the 10th, the continuance of the gale, and of the fog more opaque, if possible, than before, and more wet, were not only productive of irksome detention, but they prevented us from taking exercise; our walks being confined to a space between the marks which the Esquimaux had put up on two projecting points, whither we went at every glimpse of clearness, to examine into the state of the waves. We witnessed with regret, in these short rambles, the havoc which this dreary weather made amongst the flowers. Many that had been blooming on our arrival, were now lying prostrate and withered. These symptoms of decay could not fail painfully to remind us that the term of our operations was fast approaching; and often, at this time, did every one express a wish that we had some decked vessel, in which the provision could be secured from the injury of salt water, and the crew sheltered when they required rest, that we might quit this shallow coast, and steer at once towards Icy Cape.

The annexed sketch taken by Lieutenant Back, conveys a better picture of our encampment, and of the murkiness of the atmosphere, than any description of mine could do, and points out the propriety of designating this dreary place by the name of Foggy Island. As an instance of the illusion occasioned by the fog, I may mention that our hunters sallied forth, on more than one occasion, to fire at what they supposed to be deer, on the bank about one hundred yards from the tents, which, to their surprise, took wing, and proved to be cranes and geese.

The wind changed from east to west in the course of the night, and at eight in the morning of the 11th, the fog dispersed sufficiently to allow of our seeing a point bearing N.W. by W., about three miles and a half distant, which we supposed to be an island. We, therefore, hastened to embark; but before the boats could be dragged so far from the shore that they would float, the fog returned. The wind, however, being light, we resolved to proceed, and steer by compass, to the land that had been seen. Soon after quitting the beach we met with shoals, which forced us to alter the course more to the north; and having made the distance at which we estimated the point to be, and being ignorant which way the coast trended beyond it, we rested for some time upon the oars, in the hope that the fog would clear away, even for a short time, to enable us to shape our course anew; but in vain; all our movements in the bay being impeded by the flats that surrounded us, we were compelled to return to Foggy Island. Scarcely had the men made a fire to dry their clothes, which were thoroughly wet from wading over the flats, than the fog again dispersing, we pushed off once more. On this occasion we arrived abreast of the point whilst the weather continued clear, but found a reef, over which the waves washed, stretching to the north-west, beyond the extent of our view. Just as we began to proceed along the reef, the recurrence of

1826. August.

Friday, 11th.

the fog rendered it necessary for us to seek for shelter on the shore; and as we were heartily tired of our late encampment, we endeavoured to find another, but the shoals prevented our reaching any landing-place. We, therefore, retraced our course, though with much reluctance, to Foggy Island, which the men declared to be an enchanted island. Though our wanderings this day did not exceed seven miles, the crews were employed upwards of two hours in dragging the boats through the mud, when the temperature of the water was at 40°, and that of the air 41°. They endured this fatigue with the greatest cheerfulness, though it was evident they suffered very much from the cold; and in the evening we witnessed the ill effects of this kind of labour by finding their legs much swelled and inflamed. The fear of their becoming ill from a frequent repetition of such operations, made me resolve not to attempt the passage of these flats again till the weather should be so clear that we might ascertain their extent, and see in what way they might be passed with less risk. Fog is, of all others, the most hazardous state of the atmosphere for navigation in an icy sea, especially when it is accompanied by strong breezes, but particularly so for boats where the shore is unapproachable. If caught by a gale, a heavy swell, or drifting ice, the result must be their wreck, or the throwing the provisions overboard to lighten them, so as to proceed into shoal water. Many large pieces of ice were seen on the borders of the shallow water; and from the lowness of the temperature, we concluded that the main body was at no great distance. We had also passed through a stream of perfectly fresh water, which we supposed was poured out from a large river in the immediate vicinity, but the fog prevented our seeing its outlet. The atmosphere was equally foggy throughout the

night, and all the 12th, although the wind had changed to the east, and blew a strong breeze. Winds from this quarter had been extremely prevalent since the preceding April: but on our former visit to the Polar Sea, they had been of rare occurrence, and confined to the spring months, which we passed at Fort Enterprise. The obstinate continuance of fog forms another material difference between this season and the same period of 1821. We were only detained three times in navigating along the coast that year to the east of the Coppermine River; but on this voyage hardly a day passed after our departure from the Mackenzie that the atmosphere was not, at some time, so foggy as to hide every object more distant than four or five miles. The day that I visited Mount Conybeare, and that spent on Flaxman Island, form the only exceptions to this remark. A question, therefore, suggests itself :—Whence arises this difference? which, I presume, can be best answered by reference to the greater accumulation of ice on this coast, and to the low and very swampy nature of the land. There is a constant exhalation of moisture from the ice and swamps during the summer months, which is, perhaps, prevented from being carried off by the vicinity of the Rocky Mountains, and, therefore, becomes condensed into a fog. The coast to the eastward of the Coppermine River is high and dry, and far less encumbered with ice.

Some deer appearing near the encampment, a party was despatched in pursuit of them; but having been previously fired at by Augustus, they proved too wary. The exertions of the men were, however, rewarded by the capture of some geese and ducks. The whole of the vegetation had now assumed the autumnal tint.

There was not the least abatement in the wind, or change in the murky atmosphere, throughout the 13th. The party assembled at divine service, and afterwards amused themselves as they could in their tents, which were now so saturated with wet as to be very comfortless abodes; and in order to keep ourselves tolerably

1826.
August.

Sunday,
13th.

warm we were obliged to cover the feet with blankets; our protracted stay having caused such a great expenditure of the drift-wood, that we found it necessary to be frugal in its use, and only to light the fire when we wanted to cook the meals. The nights, too, we regretted to find, were lengthening very fast; so that from ten P.M. to two A.M., there was too little light for proceeding in any unknown track.

Monday,
14th.
  The wind this day was moderate, but the fog was more dense, and very wet. Tired, however, of the confinement of the tent, most of the party wandered out in search of amusement, though we could not see one hundred yards; and some partridges, ducks, and geese, were shot.

Tuesday,
15th.
  The fog was dispersed at seven in the morning of the 15th, by a north-east gale, which created too great a surf on the beach for us to launch the boats, and the fog returned in the evening. The temperature fell to 35°, and in the course of the night ice was formed on the small pools near the encampment. Augustus set off in the afternoon to cross over to the main shore for the purpose of hunting, and to see whether there were any traces of the western Esquimaux, but he found none, and only saw three rein-deer.

Wednesday,
16th.
  The weather again became clear, after the sun rose, on the 16th, and we embarked as soon as the flowing of the tide enabled us to launch the boats, all in the highest spirits at the prospect of escaping from this detestable island. We took advantage of the fair wind, set the sails, and steered to the westward parallel to the coast. We had never more than from three to six feet water, for the first seven miles, until we had passed round the reef that projects from the point we had so often attempted to reach, and which was named Point Anxiety.

  Between Point Anxiety and Point Chandos, which is eight miles further to the westward, the land was occasionally seen;

but after rounding the latter point we lost sight of it, and steered to the westward across the mouth of Yarborough Inlet, the soundings varying from five feet to two fathoms. The fog returned, and the wind freshening, soon created such a swell upon the flats, that it became necessary to haul further from the land; but the drift ice beginning to close around us, we could no longer proceed with safety, and, therefore, endeavoured to find a landing-place. An attempt was made at Point Heald, and another on the western point of Prudhoe bay, but both were frustrated by the shoalness of the water, and the height of the surf. The increasing violence of the gale, however, and density of the fog, rendering it absolutely necessary for us to obtain some shelter, we stood out to seaward, with the view of making fast to a large piece of ice. In our way we fell among gravelly reefs, and arriving at the same time suddenly in smooth water, we effected a landing on one of them. A temporary dispersion of the fog showed that we were surrounded with banks nearly on a level with the water, and protected to seaward by a large body of ice lying aground. The patch of gravel on which we were encamped, was about five hundred yards in circumference, destitute of water, and with no more drift wood than a few willow branches, sufficient to make one fire.

CHAPTER V.

Commence Return to the Mackenzie—Delayed again at Foggy Island—Ice packed on the
Reefs near Beaufort Bay, and on the Coast about Clarence River—Pass the Channels
near Herschel Island in a Gale and Fog—A sudden Gale—Escape an Attack which
the Mountain Indians meditated—Enter the Mackenzie—Peel River—Arrival at Fort
Franklin.

1826.
August.
Wednesday,
16th.

THE period had now arrived when it was incumbent on me to
consider, whether the prospect of our attaining the object of the
voyage was sufficiently encouraging to warrant the exposure of
the party to daily increasing risk, by continuing on. We were
now only half way from the Mackenzie River to Icy Cape;
and the chance of reaching the latter, depended on the nature
of the coast that was yet unexplored, and the portion of the
summer which yet remained for our operations.

I knew, from the descriptions of Cook and Burney, that the
shore about Icy Cape resembled that we had already passed, in
being flat, and difficult of approach ; while the general trending
of the coast from the Mackenzie to the west-north-west, nearly
in the direction of Icy Cape, combined with the information we
had collected from the Esquimaux, led me to conclude that no
material change would be found in the intermediate portion.

The preceding narrative shows the difficulties of navigating
such a coast, even during the finest part of the summer; if,
indeed, any portion of a season which had been marked by a
constant succession of fogs and gales could be called fine. No
opportunity of advancing had been let slip, after the time of our

arrival in the Arctic Sea; and the unwearied zeal and exertion of the crews had been required, for an entire month, to explore the ten degrees of longitude between Herschel Island and our present situation. I had, therefore, no reason to suppose that the ten remaining degrees could be navigated in much less time. The ice, it is true, was more broken up, and the sea around our present encampment was clear; but we had lately seen how readily the drift ice was packed upon the shoals by every breeze of wind blowing towards the land. The summer, bad as it had been, was now nearly at an end, and on this point I had the experience of the former voyage for a guide. At Point Turnagain, two degrees to the south of our present situation, the comparatively warm summer of 1821 was terminated on the 17th of August, by severe storms of wind and snow; and in the space of a fortnight afterwards, winter set in with all its severity. Last year, too, on the 18th and following days of the same month, we had a heavy gale at the mouth of the Mackenzie; and appearances did not indicate that the present season would prove more favourable. The mean temperature of the atmosphere had decreased rapidly since the sun had begun to sink below the horizon, and the thermometer had not lately shown a higher temperature than 37°. Ice, of considerable thickness, formed in the night, and the number of the flocks of geese which were hourly seen pursuing their course to the westward, showed that their autumnal flight had commenced.

While a hope remained of reaching Behring Straits, I looked upon the hazard to which we had, on several occasions, been exposed, of shipwreck on the flats, or on the ice, as inseparable from a voyage of the nature of that which we had undertaken; and if such an accident had occurred, I should have hoped, with a sufficient portion of the summer before me, to conduct my party in safety back to the Mackenzie. But the loss of the

Y

boats when we should have been far advanced, and at the end of the season, would have been fatal. The deer hasten from the coast as soon as the snow falls: no Esquimaux had been lately seen, nor any winter-houses, to denote that this part of the coast was much frequented; and if we did meet them under adverse circumstances, we could not, with safety, trust to their assistance for a supply of provision; nor do I believe that, if willing, even they would have been able to support our party for any length of time.

Till our tedious detention at Foggy Island, we had had no doubt of ultimate success; and it was with no ordinary pain that I could now bring myself even to think of relinquishing the great object of my ambition, and of disappointing the flattering confidence that had been reposed in my exertions. But I had higher duties to perform than the gratification of my own feelings; and a mature consideration of all the above matters forced me to the conclusion, that we had reached that point beyond which perseverance would be rashness, and our best efforts must be fruitless. In order to put the reader completely in possession of the motives which would have influenced me, had I been entirely a free agent, I have mentioned them without allusion to the clause in my instructions which directed me to commence my return on the 15th or 20th of August, " if, in consequence of slow progress, or other unforeseen accident, it should remain doubtful whether we should be able to reach Kotzebue's Inlet the same season."

In the evening I communicated my determination to the whole party; they received it with the good feeling that had marked their conduct throughout the voyage, and they assured me of their cheerful acquiescence in any order I should give. The readiness with which they would have prosecuted the voyage, had it been advisable to do so, was the more creditable, because many of them had their legs swelled and inflamed from

1826.
August.

continually wading in ice-cold water while launching the boats, not only when we accidentally ran on shore, but every time that it was requisite to embark, or to land upon this shallow coast. Nor were these symptoms to be overlooked in coming to a determination ; for though no one who knows the resolute disposition of British sailors can be surprised at their more than readiness to proceed, I felt that it was my business to judge of their capability of so doing, and not to allow myself to be seduced by their ardour, however honourable to them, and cheering to me.

Compelled as I was to come to the determination of returning, it is a great satisfaction to me to know, as I now do, that the reasons which induced me to take this step were well-founded. This will appear by the following extract from Captain Beechey's official account of his proceedings in advancing eastward from Icy Cape, with which I have been favoured.

" Mr. Elson, (the master,) after quitting the ship off Icy Cape, on the 18th August, had proceeded along the coast without interruption, until the 22nd of the month, when he arrived off a very low sandy spit, beyond which, to the eastward, the coast formed a bay, with a more easterly trending than that on the west side ; but it was so low that it could not be traced far, and became blended with the ice before it reached the horizon. It was found impossible to proceed round the spit, in consequence of the ice being grounded upon it, and extending to the horizon in every direction, except that by which the boat had advanced, and was so compact that no openings were seen in any part of it. This point, which is the most northern part of the continent yet known, lies in latitude, by meridian altitude of the sun, 71° 23′ 39″ N. ; and longitude, by several sets of lunar distances, both observed on an iceberg, 156° 21′ W. ; and is situated one hundred and twenty miles beyond Icy Cape. Between these

Y 2

two stations, and, indeed, to the southward of the latter, the coast is very flat, abounding in lakes and rivers, which are too shallow to be entered by any thing but a baidar. The greater part of the coast is thickly inhabited by Esquimaux, who have their winter-habitations close to the beach.

" The barge had not been off this point sufficiently long to complete the necessary observations, when the same westerly wind, which had induced me to proceed round Cape Lisburn, brought the ice down upon the coast, and left the boat no retreat. It at the same time occasioned a current along shore to the northward, at the rate of three and four miles per hour. The body of ice took the ground in six and seven fathoms water, but pieces of a lighter draft filled up the space between it and the shore, and, hurried along by the impetuosity of the current, drove the barge ashore, but fortunately without staving her. By the exertions of her officers and crew she was extricated from this perilous situation, and attempts were made to track her along the land wherever openings occurred, in execution of which the greatest fatigue was endured by all her crew. At length all efforts proving ineffectual, and the spaces between the ice and the shore becoming frozen over, it was proposed to abandon the boat, and the crew to make their way along the coast to Kotzebue Sound, before the season should be too far advanced. Preparations were accordingly made; and that the boat might not be irrevocably lost to the ship, it was determined to get her into one of the lakes, and there sink her, that the natives might not break her up, and from which she might be extricated the following summer, should the ship return. During this period of their difficulties they received much assistance from the natives, who, for a little tobacco, put their hands to the tow-rope. Their conduct had, in the first instance, been suspicious; but in the time of their greatest distress, they were well disposed, bringing venison, seal's flesh, oil, &c., and offered up a prayer that the wind would blow off the shore, and liberate the boat from her critical situation.   Before the necessary arrangements were made respecting the barge, appearances took a more favourable turn ; the ice began to move off shore, and after much tracking, &c., the boat was got clear, and made the best of her way toward the sound ; but off Cape Lisburn she met with a gale of wind, which blew in eddies so violently, that it is said the spray was carried up to the tops of the mountains ; and the boat, during this trial, behaved so well, that not a moment's anxiety for her safety was entertained.   I must not close the account without expressing my warm approbation of the conduct of Mr. Elson."

The barge rejoined Captain Beechey on the 10th September, at Chamisso Island, the Blossom having gone thither to wood and water, and being further forced to quit the coast to the northward, in consequence of strong westerly winds.

Could I have known, or by possibility imagined, that a party from the Blossom had been at the distance of only one hundred and sixty miles from me, no difficulties, dangers, or discouraging circumstances, should have prevailed on me to return ; but taking into account the uncertainty of all voyages in a sea obstructed by ice, I had no right to expect that the Blossom had advanced beyond Kotzebue Inlet, or that any party from her had doubled Icy Cape.   It is useless now to speculate on the probable result of a proceeding which did not take place ; but I may observe, that, had we gone forward as soon as the weather permitted, namely, on the 18th, it is scarcely possible that any change of circumstances could have enabled us to overtake the Blossom's barge*.

* I have recently learned, by letter from Captain Beechey, that the barge turned back on the 25th of August, having been several days beset by the ice.   He likewise informs me, that the summer of 1827 was so unfavourable for the navigation of the

The wind changed to the N.E. after midnight, the squalls were more violent, and in the morning of the 17th such a surf was beating on the borders of the reef, that the boats could not be launched. The fog disappeared before the gale about eleven, and during the afternoon we enjoyed the clearest atmosphere that we had witnessed since our departure from Mount Conybeare. This was the first opportunity there had been, for the seven preceding days, of making astronomical observations, and we gladly took advantage of it, to observe the latitude, 70° 26′ N.; longitude 148° 52′ W.; and variation 41° 20′ E. We had likewise the gratification of being able to trace the land round Gwydyr Bay, to its outer point, bearing S. 79 W. ten miles, which I have named after my excellent companion Lieutenant Back, and of seeing a still more westerly hummock, bearing S. 84 W., about fifteen miles, that has been distinguished by the name of my friend Captain Beechey; at which point, in latitude 70° 24′ N., longitude 149° 37′ W., our discoveries terminated. The fog returned at sunset, and as the wind was piercingly cold, and we had neither fire nor room for exercise, we crept between the blankets, as the only means of keeping ourselves warm.

The gale having considerably abated, and the weather being clear, we quitted Return Reef on the morning of the 18th, and began to retrace our way towards the Mackenzie. As the waves were still very high to seaward, we attempted to proceed inside of the reefs, but as the boats were constantly taking the ground, we availed ourselves of the first channel that was sufficiently deep to pull on the outside of them. The swell being too great there for the use of the oars, the sails were set double reefed,

northern coast of America, that the Blossom did not reach so high a latitude as in the preceding year; nor could his boat get so far to the east of Icy Cape, by one hundred miles. The natives, he says, were numerous, and, in some instances, ill-disposed.

and the boats beat to the eastward against the wind, between the drift ice and the shallow water.

A gale rose after noon from N.E. by N., which enabled us to shape a course for Foggy Island, where we arrived at three P.M., just at a time when the violence of the squalls, and the increased height of the swell, would have rendered further proceeding very hazardous. We now enjoyed the comforts of a good fire and a warm meal, which we had not had since the evening of the 16th. The men were afterwards employed in erecting a square pile of drift timber, on the highest part of the island fronting the sea, on which a red cornet flag was left flying, and underneath it was deposited, in a tin case, a letter for Captain Parry, containing an account of our proceedings; also a silver medal and a halfpenny: and in order that government might have some chance of hearing of our proceedings, should any accident subsequently befal the party, there was also deposited an unsealed letter, wrapped in bark, addressed to the Russian Fur Traders, in the expectation that the Esquimaux might probably convey it to their Establishment. An ice chisel, a knife, a file, and a hatchet, were hung up on the pile, for the Esquimaux. On digging to erect these posts, the ground was found frozen at the depth of sixteen inches; and the thermometer, during the day, seldom rose above 37°. This evening the temperature was 33°. We were vexatiously detained the 19th, and following day, by the continuance of the gale, and a thick fog; during which time many large flocks of geese were observed passing away to the westward. The tides were now much higher than during our first visit.

The breeze was moderate on the morning of the 21st, yet we were prevented from embarking until ten o'clock, by the return of the fog. We then hastened to escape from this ill-omened island. The boats were pulled to seaward, so as to gain a sufficient offing for them to pass on the outside of the shallow water; and by the aid of the oars and sails we made good progress, and encamped within sight of Flaxman Island. A black whale, a seal of the largest kind, and numerous flocks of geese, were seen in the course of this day. Several stars were visible after ten P.M. Showers of snow fell during the night, but the morning of the 22nd was calm and clear. We embarked at day-light, and in the course of three hours arrived abreast of the east end of Flaxman Island. The ice had broken from the northern shore during our absence, and was now lying about a mile from the land, apparently aground on reefs, as we had observed it to be along the outer border of the one at the west end of the island. The water was much deeper between Flaxman Island and the main, than when we passed in the early part of August. Eastward of Point Brownlow there was an open channel of three or four miles wide, and by keeping close to the borders of the drift ice we avoided the shallows at the mouth of the Canning River, and arrived at Boulder Island about noon. Here we found an Esquimaux grave, containing three bodies, covered with drift timber, and by their side there were placed the canoes, arrows, and fishing implements of the deceased. Not being able to procure fresh water here, we set forward to cross Camden Bay, touched at one of the points to fill the water-casks, and reached Barter Island after dark; the crews much fatigued, having been pulling for nineteen hours. We regretted to find the Esquimaux had visited this spot during our absence, and carried away the gun and ammunition which had been left by mistake at the encampment on the 4th of August, because we were not only apprehensive that some accident might have occurred in the attempt to discharge the gun, but were very desirous to prevent the introduction of fire-arms among these people. Being now near the point of the coast at which we had seen a considerable

number of the natives, we remained at the encampment until ten
o'clock on the morning of the 23rd, to clean the guns and issue a fresh supply of ammunition to the party. The day was calm and cloudless; the whole range of the Romanzoff Mountains was in sight, and they appeared to be more covered with snow than when we passed to the westward. A few musquitoes made their appearance, but they were very feeble. Having landed at Point Manning to replenish the water-casks, we afterwards pulled throughout the day close to the edge of the ice, which was still heavy, though loose, and encamped near Point Griffin. Some large-sized medusæ, and several of the gelatinous substances known to seamen by the name of blubber, were found on the beach, which accounted for the number of black ducks that had been seen in the course of the day, as they feed on those substances. The temperature varied this day from 35° to 46°; and the thermometer rose to 64° at two P.M., when exposed to the sun's rays.

The morning of the 24th was calm; we set forward at daylight,
and having proceeded a few miles between heavy floating ice, about half a mile from the shore we met with a large sheet of bay ice of last night's formation, of sufficient thickness to impede though not to stop the boats. Having arrived abreast of Point Humphreys, we steered out to seaward, for the purpose of avoiding the shallows that extend across Beaufort Bay, intending to direct our course in a line for Mount Conybeare, which was in sight. We were then exposed to a long rolling swell, and we soon afterwards perceived that it had driven the ice upon the reefs at the eastern extremity of the bay, which would have precluded our retreat to the shore in the event of the wind rising. It therefore became necessary to penetrate into the pack, and keep by the side of the reefs; but in doing so, the boats were exposed to no little danger of being broken in passing through

1826.
August.

the narrow channels between the masses of ice which were tossing with the swell, and from which large pieces frequently fell. At six P.M. we passed our former encampment on Icy Reef, and afterwards proceeded through an open space to Demarcation Point, where we encamped, and hauled up the boats to prevent them from being injured by the surf. We found here two families of Esquimaux, which belonged to the party that had been to Barter Island, waiting the return of a man from hunting, in order to follow their companions to the eastward. They showed much joy at seeing us again, and remained the greater part of the night talking with Augustus. The most active young man of the party, not thinking himself sufficiently smart for the occasion, retired to the oomiack to change his dress and mouth ornaments, capering about on his return, evidently proud of his gayer appearance.

Friday,
25th.

The morning was foggy, but there being little wind, we launched the boats, and pulled for an hour close to the shore, when we came to a body of ice so closely packed as scarcely to afford a passage, and it was with difficulty that we arrived at Clarence River. There we perceived four tents; near which we had been warned by our visiters last night not to land, as the party had recently lost their parents, and it was feared that, in the state of mind in which they then were, they might be disposed to do us some injury. We pulled near enough to inquire about the gun, and learned that the person who had it was farther to the eastward. The difficulties of forcing a passage were not diminished beyond this place, and we were further impeded in our advance by new ice formed between the larger masses, which required additional labour to break through. The fog cleared away at ten; we halted to breakfast at Backhouse River, and remained whilst Augustus went in pursuit of two rein-deer, one of which he killed.

Renewing our course, we passed on the outside of the ice until we were nearly abreast of Mount Conybeare, when the wind came strong from the eastward, and obliged us to have recourse again to its shelter. This barrier, however, terminated at the end of five miles, and being then exposed to the wind and swell, against which the men were unable to pull, we encamped.

The experience we had now gained of the ice being packed upon this shore by a wind from the sea, assured us of the correctness of the report which the Esquimaux had given, and likewise afforded a reason for their expression of surprise at our being unprovided with sledges, as it was evident, unless a strong wind blew from the land, that the new ice would soon unite the pack with the shore, and preclude the possibility of making the passage in boats, unless by going outside of the ice, which would be extremely hazardous, from the want of shelter in the event of a gale springing up. The pieces of ice were generally from ten to fifteen feet in height, many of them were from twenty to thirty feet: their length was from twenty to one hundred yards. We saw several white whales in the open water, and a flock of white geese at the encampment, which were the first noticed on this coast. The rising of the wind from E.N.E. this afternoon was accompanied by an increase of temperature from 43° to 53°, and we felt a comfortable sensation of warmth, to which we had been strangers for the preceding month.

We took advantage of a favourable breeze to embark before daylight on the morning of the 26th; at sunrise it increased to a gale, and raised a heavy sea. In two hours we ran to the commencement of the intricate channels leading to Herschel Island, where the Esquimaux seen at Barter Island were encamped on a reef, and apparently gazing in astonishment at the speed of our boats. They made many signs for us to land, which we were desirous of doing had it been practicable for the surf. That the boats might be perfectly manageable, we took two reefs in the sails, and shaped the course for Herschel Island; but scarcely were the sails reset before a fog came on that hid every mark that could guide us; a heavy swell was rolling at the time, and to arrive at the island we had to pass through a channel only about two hundred yards broad. To find this, surrounded as it was by shoals, in the midst of a dense fog, was a task of considerable anxiety and danger, and our situation was not rendered more agreeable by being assailed the whole way with continued shouting from persons to us invisible; our arrival having been communicated by the Esquimaux who first descried us, to their companions on the neighbouring reefs. We effected it, however, and landed in safety, though we did not discover the island till we were within forty yards of its shore. We had scarcely landed before the fog dispersed, and discovered to us a solitary tent on an adjacent point. Three men soon paid us a visit, whom we had not seen before, and they informed us that nearly the whole of the tribe was now collected in the vicinity for the purpose of hunting deer, and catching whales and seals for the winter's consumption. We quitted the island at ten A.M., and steered directly for Point Kay, to avoid the sinuosities of the coast, and the frequent interruption of the Esquimaux, whose tents were observed to be scattered on the beach nearly the whole way to Babbage River. Three men and some women came off to bring us fish, and being liberally rewarded, they went away perfectly happy, singing the praises of the white people. We passed round Point Kay at four P.M., with a moderate breeze from W.N.W., and steered for Point King, keeping about two miles from the land. As the afternoon wore away, gloomy clouds gathered in the north-west; and at six a violent squall came from that quarter, attended with snow and sleet. The gale increased with rapidity: in less than ten minutes the sea was white with foam, and such waves were raised as I had never before been exposed to in a boat. The spray and sea broke over us incessantly, and it was with difficulty that we could keep free by baling. Our little vessels went through the water with great velocity under a close-reefed sail, hoisted about three feet up the main-mast, and proved themselves to be very buoyant. Their small size, however, and the nature of their construction, necessarily adapted for the navigation of shallow rivers, unfitting them for withstanding the sea then running, we were in imminent danger of foundering. I therefore resolved on making for the shore, as the only means of saving the party, although I was aware that, in so doing, I incurred the hazard of staving the boats, there being few places on this part of the coast where there was sufficient beach under the broken cliffs. The wind blowing along the land we could not venture on exposing the boat's side to the sea by hauling directly in, but, edging away with the wind on the quarter, we most providentially took the ground in a favourable spot. The boats were instantly filled with the surf, but they were unloaded and dragged up without having sustained any material damage. Impressed with a sense of gratitude for the signal deliverance we had experienced on this and other occasions, we assembled in the evening to offer up praise and thanksgiving to the Almighty.

On the 27th the weather was calm; but as a heavy surf prevented our embarkation, we took advantage of the delay to dry our bedding, clothes, and pemmican. The guns were likewise cleaned, and every thing put in proper order. There was an Esquimaux party at this spot, which had witnessed the landing of the boats in the storm with astonishment, having expected to see every man drowned. Augustus passed the night at their tents; and having brought the whole party to our encampment, the women, with much good nature, sewed soles of seal-skins to

1826.
August.

Sunday,
27th.

the men's mocassins, in order to fit them better for the operation of tracking, in which they were soon to be employed. These Esquimaux had recently returned from a visit to the gang that had pillaged the boats at the mouth of the Mackenzie; and we now learned the intention that had been entertained of destroying our party, along with the other particulars that have been already mentioned.

Our approach to the Mackenzie was marked by the quantity of drift timber floating about. We passed several families of the natives, without visiting them, until we perceived one party taking some fish from their nets, which tempted us to land. The fish were large *tittameg* and *inconnû*, and proved remarkably fine. We again embarked, but having to pull head to the sea, we took in much water, and were glad to seek shelter on a gravel reef, where three Esquimaux tents were pitched. The whole party quickly mustered around us, and we were not a little surprised to find so many inhabitants as twenty-seven, including women and children, in three tents only; but on inquiry we found that the number was not greater than usual. Two of the men were very aged and feeble, the rest were young and active. They practise jumping, as an amusement, from their youth; and we had an opportunity of witnessing some of their feats, which displayed much agility. The women cheerfully repaired our mocassins, and their industry, as well as the good conduct of the men, were rewarded by some valuable presents. We were astonished to learn that there had been fog only a day and a half in this neighbourhood since we passed, but the wind had been generally strong. Augustus gained some information respecting the western Esquimaux, and the coast to the westward, which he did not communicate to me until some days afterwards, otherwise I should have endeavoured to elicit more satisfactory details. It was to the following purport:—The western Esquimaux having purchased the furs from those men that dwell near the Macken-zie, at Barter Island, proceed to the westward again without delay. A few days journey beyond a part of the coast which Augustus understood from description to be Return Reef, the sea is still more shallow than that which we had navigated, and the water is still, except at certain periods of the year, when it is agitated like a strong rapid, by the efflux of the waters of a deep inlet, or strait. The land is visible on both sides from the middle of this opening ; the Esquimaux make for the west side, and on reaching it relinquish their canoes, and drag their furs overland to the Russian establishments, which are situated in the interior, where the land is more elevated than on the coast. The mountain Indians come down annually in large parties to this inlet, and warm contests often arise between them and the Esquimaux. The latter are frequently worsted, from their inferior numbers, and lose their property, which the Indians bring by land to the neighbourhood of Herschel Island, to dis-pose of to the Esquimaux in that quarter. The direction of the inlet was supposed, by Augustus, to be about south-west. I am inclined to think that it is the estuary of a large river, flowing to the west of the Rocky Mountains, obstructed by sand-banks, like the mouth of the Mackenzie. In the course of the day three Esquimaux, who had seen our tent from a distance, came to visit us. One of them was recognised to have been of the party which attacked us at the mouth of the Mackenzie. He gave Augustus a detailed account of their schemes on that occasion, which exactly corresponded with that we had received on the preceding day. He further told us that the party which had assailed us had certainly removed to the eastward ; but if any of them should have remained, to watch our motions, they could be avoided by entering the river by a more westerly branch than the one which we had descended, and offered to guide us thither.

This man was very intelligent, and having carefully examined the boats, intimated that he would construct an oomiack after the same plan. We embarked at four in the evening with our new friend for a guide, and in a short time arrived at the main shore where his tent stood, and where he asked the party to encamp, as he intended to go no farther. We were not, however, so disposed; and having filled the casks with fresh water, and made some presents to the women, we pushed off to take advantage of the remaining daylight in getting round a reef which projected far to seaward. We could not effect this, and at sunset, not being able to land on the reef on account of the shallowness of the water, we put back to within a mile and a half of the Esquimaux tents. Garry Island was seen soon after sunset; and the aurora borealis appeared in the night for the first time this season. The temperature of the air varied from 30° to 49°, and that of the sea water was 37° 2′, a quarter of a mile

from the shore. A gale coming on in the night, and continuing till the following evening, detained us on shore. During our stay we were visited by a numerous party of Esquimaux, and found it necessary to draw a line round the tents, which they were not permitted to pass. These people told us that Dr. Richardson's party had been seen clear of the Mackenzie, and had given kettles to men in three canoes, after escaping an attempt made by the Esquimaux to drag the boats on shore. This account, showing that the propensity to plunder was not confined to the Esquimaux with whom we had met, excited painful apprehensions for the safety of the eastern party, if they should find it necessary to return by the Mackenzie, because we now learned that the natives collect in numbers near its mouth at the close of summer. In ordinary seasons the weather is mild, and the winds variable until the ice breaks up, which is usually about the end of August, when north-west winds, and stormy weather, are expected. In this season, however, the winds had been so boisterous that the Esquimaux had seldom been able to venture out to sea, and their whale fishery had consequently failed. Our visitors left us about two P.M.; but, shortly afterwards, we heard loud cries, and on looking round saw two young Esquimaux running in breathless haste to announce that a large party of Indians had come down from the mountains with the express purpose of attacking the boats and killing every man of the party. They desired us to embark instantly, as the only means of escape; for the Indians, they said, were already at the tents within our view, and when they left them they were on the point of spreading round us to commence the onset. They further said, that the Indians, having been provoked by our trading with the Esquimaux, had been along the coast in search of us, and that it was only this afternoon they had espied our tents, which, by their fluttering in the wind, they knew did not belong to the Esquimaux. On this discovery they had come to the nearest party of Esquimaux to make known their intention, and to request their aid. They were met by our two young friends, who were out hunting, but who returned with them to their tents, and after learning the plans in agitation, had stolen off to apprise us of our danger. As soon as Spinks returned, who had gone to shoot, we shoved off; and never were men more delighted than our two Esquimaux friends seemed to be at our escape; and especially at that of Augustus, to save whom, they asserted more than once, was their principal motive in coming to us. While Spinks was out of sight, they climbed up to the top of an old house to look for him, with the greatest apparent solicitude, and were the first to discover him returning. Up to the time of his arrival they kept repeating every particular respecting the Indians, and pointing out the mode of avoiding them. It was their intention, they

2 A

said, to pursue us to the Mackenzie, but that we should get there before them, because there were two rivers in the way which the Indians would have great difficulty in crossing, being unprovided with canoes. They urged us to make all speed, and not to halt in the night, nor to go to sleep; but, if the crew became tired, to put up on an island out of gun-shot of the main shore, because the Indians were armed with guns as well as bows. They instructed Augustus minutely as to the course we were to steer round the reef, and directed us to keep along the main shore until we should come to a large opening, which was the western outlet of the Mackenzie, and had a deep channel. We rewarded their friendly conduct by a considerable present of iron, which they received with an indifference that shewed them not to have been actuated by interested motives in making the communications. Previous to the arrival of these men we had perceived the smoke of a distant fire, which we had little regarded, supposing it to have been made by some Esquimaux who were hunting, but which, it seems, was the fire of the Indians. Having pulled round the reef, and being aided by a westerly breeze, we soon regained the main shore, and passed the mouth of the two rivers of which the Esquimaux had spoken. The night beginning to close in we pulled up to the head of an inlet; when

Wednesday, 30th. heavy rain and squalls coming on, we determined to halt. As soon as the day dawned, which was about half-past two in the morning, we returned to Shoal Water Bay; and, sailing along the coast for two or three miles to the eastward, arrived at another opening, in which the water was fresh, and we did not doubt but it would prove the deep channel by which we had been instructed to ascend. There was plenty of water near its mouth, but it gradually shoaled; and, at the distance of four miles, we ascertained that this promising opening was likewise an inlet. I now relinquished the search for a more westerly outlet than the one by which we had descended, and, therefore, steered for Pillage Point, which soon afterwards came in sight. After dragging the boats for two hours over the shoals, we rounded Pillage Point at ten A.M., and reached the deep water most opportunely; for, almost at the instant, a violent north-west gale came on, attended by thunder, lightning, and torrents of rain. The wind, however, was fair, and brought so much water into the channel of the river, that we passed, without obstruction, the shallow parts above Pillage Point. A temporary cessation of the rain at noon enabled us to land to breakfast; and we afterwards continued to scud before the gale until sunset, when we encamped. The temperature fell from 48° to 40° in the gale, and we had several showers of snow.

During the above run Augustus entertained us with an account, which he had learned from the two Esquimaux, respecting the Mountain Indians; the substance was as follows:—Seven men of that tribe had been to Herschel Island to trade with the Esquimaux, who shewed them the different articles they had received from us, and informed them of our being still on the coast, and that our return by this route was not improbable. This intelligence they set off at once to communicate to the rest of their tribe, who, supposing that we should ruin their trade with the Esquimaux, resolved on coming down in a body to destroy us; and that they might travel with expedition, their wives and families were left behind. They came to the sea coast by the Mountain Indian River, opposite Herschel Island, and finding that we had not returned, but supposing it possible that we might pass them there, as they had no canoes to intercept us, they determined on travelling to the mouth of the Mackenzie, where they could conveniently subsist by fishing and hunting until our arrival. They had been informed of the manner in which we had been robbed by the Esquimaux at that place, and they formed a similar plan of operations. When our crews were wading and launching the boats over the flats in Shoal Water Bay, a few of them were to have offered their assistance, which they imagined would be readily accepted, as we should probably take them for Indians belonging to the Loucheux tribe, with whom we were acquainted. While pretending to aid us they were to have watched an opportunity of staving the boats, so as to prevent them from floating in the deeper channel, which runs close to the land near Pillage Point. The rest of the party, on a signal being given, were then to rush forth from their concealment, and join in the assault. They were, in pursuance of this plan, travelling towards the Mackenzie, when they discovered our tents; and it appeared that the two young men who brought us the intelligence, had been sent as an act of gratitude by an old Esquimaux, to whom we had given a knife and some other things, on the preceding day. After hearing the plans of the Indians, he called the young men aside and said to them, " These white people have been kind to us, and they are few in number, why should we suffer them to be killed ? you are active young men, run and tell them to depart instantly." The messengers suggested that we had guns, and could defend ourselves. " True," said he, " against a small force, but not against so large a body of Indians as this, who are likewise armed with guns, and who will crawl under cover of the drift timber, so as to surround them before they are aware ; run, therefore, and tell them not to lose a moment in getting away, and to be careful to avoid the flats at the mouth of the river by entering the western channel."

As the goods which the Mountain Indians exchange with the Esquimaux at Herschel Island, are very unlike those issued from any of the Hudson's Bay Company's post, I conclude that they obtain them from the Russians ; but the traders of that nation being prohibited by their government from supplying guns to any Indians, I am at a loss to account for these people having them;—perhaps, the prohibition only applies to the Esquimaux, or the people on the sea coast. That the Mountain Indians have fire-arms we learned, not only on the present occasion, but in our first interview with the Esquimaux at Herschel Island.

The few general remarks which I have to offer, on the subject of a North-West Passage, will appear in a subsequent part of the narrative; and here I shall only state, that we traced the coast, westward from the mouth of the Mackenzie, three hundred and seventy-four miles, without having found one harbour in which a ship could find shelter.

On the 31st, we continued the ascent of the river, and encamped in the evening within the limit of the spruce fir trees.

Favoured by a strong north-west gale, on the 1st of September, we sailed the whole day along the western main shore, and, generally, within view of the Rocky Mountains. One of the numerous bends of the river took us within eight miles of part of the mountains, which appeared to be composed of a yellow stone, and was from eight hundred to a thousand feet in height. In the course of the day, we came to the most northerly poplars, where the foliage had now assumed the yellow autumnal hue. The gale continued with strong squalls on the 2nd, and we advanced rapidly under double-reefed sails, though the course of the river was very winding. The temperature of the air varied from 41° to 35°. On the third we had calm weather, and still keeping the western land aboard, we were led into a river which we had not discovered in our descent. The course of this river was, for a time, parallel to our route, and we took it at first for one of the channels of the Mackenzie; but, in the afternoon, we saw a mountain to the eastward, and ascertained that we were to the southward of Point Separation.   We, therefore, began to descend the river again, and encamped shortly after sunset. Just after it became dark, voices were heard on the opposite side of the river, to which we replied, and soon afterwards, three Indians were observed crossing towards us in canoes. They approached cautiously, but on being invited to land, they did so, though one of them was so great a cripple as to require being carried from the canoe to the fire-side.   The alarm these poor people had felt, was soon dissipated by kind treatment.   They were armed with bows and arrows only, and clothed in hare skins and leather.   Their trowsers were similar to those worn by the lower Loucheux, to which tribe they, probably, belonged. We could communicate with them only by signs, except by using a few words of Chipewyan, which one of them appeared to understand.   We collected from them that they knew of Fort Good Hope, but none of them seemed to have visited it, as they had not a single article of European manufacture about their persons.   They delineated on a stone the course of the Macken-zie, and of the river we had newly discovered, which appears to flow from the Rocky Mountains, and to break through the same ridge of hill that the Mackenzie does at the Narrows.   It is probable, that it was to this river the Loucheux alluded, when they told Sir Alexander Mackenzie, opposite the present site of Fort Good Hope, that there was a river which conducted them to the sea in five days.   I have distinguished this river by the name of Peel, in honour of His Majesty's Secretary of State for the Home Department.   It is from a quarter to half a mile wide, and its banks are clothed with spruce, birch, and poplar trees, like those of the Mackenzie in the same parallel.   We set

forward at four A.M. on the 4th, with a strong favourable breeze, and in an hour, passed another river descending from the Rocky Mountains, and nearly as large as the Peel, into which it

flows. We regained the Mackenzie at noon, and at five P.M.
arrived at Point Separation, where we encamped. Here we
found the boat, rope, and kettle, in the same state in which we
had deposited them. The kettle was a great acquisition to us,
because we had suffered much inconvenience in having only one
for cooking, after the Esquimaux had robbed us of the others.
The temperature varied during the day from 29° to 55°, and, in
the evening, the sand flies were troublesome. We quitted our
encampment at day light on the 5th, and crossed the river to
look for a mark which Dr. Richardson was to have erected, if he
returned by the Mackenzie; but not finding any, we deposited a
letter and a bag of pemmican, in case he should come at a later
period, and that his party should be in want of provision. In
the vicinity of the Red River, we met Barbue, the Chief of the
Loucheux, and two or three families, who seemed in a sorry
condition from want of food, the water being too low for fishing.
The chief appeared very anxious to communicate some intelligence, which he evidently considered important, but we could
not understand him. We learned afterwards at the fort, that
it related to the death of a chief by violence on the sea coast ;
this had given rise to a rumour of the death of myself, and
afterwards of Dr. Richardson, which occasioned us, for a time,
much anxiety. The weather, on this and several days, was
remarkably fine ; berries of various kinds were very abundant
on the banks, and quite ripe. By the aid of the tracking line,
with the occasional use of the oars and sails, we proceeded up the
river at a quick rate, and reached Fort Good Hope, at half-past
four on the 7th. In consequence of the above-mentioned
rumour, I requested Mr. Bell, the gentleman in charge of the
fort, to despatch two of the Loucheux as quickly as possible to
the eastern mouth of the river, in order to gain any information
the Esquimaux could give regarding Dr. Richardson's party ; and,

that the messengers might not be delayed by hunting on the way, I left a bag of pemmican for their use. We were sorry to learn that there was some apprehension of a serious quarrel arising between the upper and lower Loucheux, in consequence of one of Barbue's sons having killed his wife, a woman of the latter tribe.

We quitted Fort Good Hope at noon on the 8th, arrived at the entrance of Bear Lake River on the 16th, and on the 21st reached Fort Franklin, where we had the happiness of meeting our friends in safety. The eastern detachment had arrived on the 1st of September, after a most successful voyage; and Dr. Richardson being anxious to extend his geological researches, as far as the season would permit, had gone in a canoe to the Great Slave Lake, having previously sent a report of his proceedings, to meet me at Fort Good Hope, in case of our being obliged to return by the Mackenzie; but the bearer of them passed us without being seen. Having read Mr. Kendall's journal, I drew up a brief account of the proceedings of both parties for the information of His Majesty's Government, and transmitted it by canoe, to Slave Lake on the following morning.

The distance travelled in the three months of our absence from Fort Franklin, amounted to two thousand and forty-eight statute miles, of which six hundred and ten were through parts not previously discovered.

I cannot close this account of our sea voyage without expressing the deep obligation I feel to Lieutenant Back for his cordial co-operation, and for his zealous and unwearied assiduity during its progress. Beside the daily delineation of the coast in the field book, the service is indebted to him for numerous drawings of scenery, as well as of the natives; and for an interesting collection of plants. My warmest thanks are likewise due to the men of my party, who met every obstacle with an ardent desire to surmount it, and cheerfully exerted themselves to the utmost of their power. Their cool, steady conduct is the more commendable, as the sea navigation was entirely novel to the whole, except the seamen Duncan and Spinks, and Hallom, the corporal of Marines. The Canadian voyagers, Felix and Vivier, first saw the ocean on this occasion.

The following Chapters contain the narrative of the proceedings of Dr. Richardson in his own words; and I embrace this opportunity of conveying my sincere thanks to him, to Mr. Kendall, and to their respective crews. I may be allowed to bear my testimony to the union of caution, talent, and enterprise in the former, which enabled him to conduct, with singular success, an arduous service of a kind so foreign from his profession and ordinary pursuits; and to the science and skill, combined with activity, of Mr. (now Lieutenant) Kendall, which must heighten the character he has already obtained for general ability and energy in his profession. I must not omit to state, that these officers describe the conduct of their crews to have been excellent.

ABSTRACT of the Mean Temperature for each Day during the Voyage along the Sea Coast west of the Mackenzie, and on the return to Bear Lake.

1826 Date.	JULY. Daily Mean.	JULY. Wind and Weather.	JULY. Situation.	1826 Date.	AUGUST. Daily Mean.	AUGUST. Wind and Weather.	AUGUST. Situation.	1826 Date.	SEPTEMBER. Daily Mean.	SEPTEMBER. Wind and Weather.	SEPTEMBER. Situation.
1	52.8	NNW, WNW moderate, gloomy	Fort Good Hope	1	42.0	NE, gale, foggy		1	38.3	NW, gale, snow	
2	58.3	West, fresh, clear		2	44.6	ENE, strong, moderate, clear	Between Icy Reef and Flaxman's Island, latitude 69° 44' and 70° 11' N., longitude 141° 30' & 145° 50' W.	2	38.6	NW, strong, clear	
3	50.3	WNW, fresh, clear	Mackenzie River, between latitude 67° 58', and 68° 53' N. longitude 130½° & 136½° W.	3	44.1	ENE, moderate, clear		3	41.1	Calm, moderate, SE, clear	
4	55.8	West, SSW, N, light, gloomy		4	40.7	East, moderate, clear		4	41.3	SE, NW, moderate, clear	
5	53.7	SW, NE, modt. gloomy, foggy		5	42.6	Calm, WbyN, moderate		5	45.9	SE, light, clear	
6	45.1	NNW, ENE, fresh, modt. rain		6	43.2	Calm, ESE, light, clear		6	51.0	Variable, light, clear	
7	41.6	SE, moderate, clear		7	42.8	ENE, fresh, clear		7	44.8	SE, light, NW. strong	
8	Not registered, Thermometer stolen by Esquimaux.		Mouth of the Mackenzie	8	42.9	ENE, strong, fog		8	41.0	NW, strong, snow	
9		ENE, strong, fog and rain		9	41.6	NE, strong, fog		9	39.3	East, moderate, clear	
10				10	39.5	ENE, strong, fog	Foggy Island, latitude 70° 16' N., longitude 147° 38' W.	10	45.8	SE, light, clear	Mackenzie River.
11			Between the Mackenzie and Herschel Island, latitude 68° 53' and 69° 34' N., longitude 136° 19' and 139° 5' W.	11	41.1	NE, moderate, fog		11	45.8	NW, moderate, rain	
12	51.6	EbyN, fresh, gloomy		12	41.1	East, moderate, very foggy		12	37.3	NW, moderate, gloomy	
13	53.3	Variable, fog and rain.		13	41.6	NE, strong, foggy		13	37.2	Calm, SE, light, clear	
14	50.5	Calm, rain, ENE, moderate, clear		14	41.3	ENE, NE, modt. foggy		14	37.9	ESE, moderate, clear	
15	48.6	Calm, clear, NW, moderate, foggy		15	38.1	NE, fresh, hazy		15	42.7	Calm, moderate, fresh, gloomy	
16	47.3	SSE, moderate, snow, fog		16	35.0	ENE, fresh, foggy	Return Reef, latitude 70° 26' N., long. 148° 52' W.	16	44.5	Variable, light, gloomy	
17	44.8	NW. North, moderate, hazy	Herschel Island	17	37.4	NE, gale, very foggy		17	36.9	Variable, moderate, rain	
18	43.6	NW, moderate, clear		18	36.2	NE, strong, clear		18	29.4	NW, fresh, gloomy	
19	43.4	NW, modt. heavy rain and fog		19	36.4	NE, strong, foggy	Foggy Island.	19	24.6	NW, moderate, gloomy	
20	39.3	NW, fresh, fog		20	36.4	NE, fresh, foggy		20	29.2	ESE, fresh, clear	
21	51.3	East, SE, clear	Between Icy Reef and Herschel Island, latitude 69° 34' and 69° 44' N., longitude 139° 5' and 141° 30' W.	21	35.7	NNE, North, moderate, clear		21	31.1	ENE, fresh, clear	Fort Franklin.
22	58.5	SE, light, clear		22	37.6	North, NE, light, clear	Between Foggy Island and the Mouth of the Mackenzie, latitude 70° 16' and 69° 53' N., Longitude 147° 38' and 136° 19' W.	Mean	39.22		
23	51.6	West, calm, East, clear		23	41.0	Calm, clear					
24	45.6	Calm, variable, clear		24	39.4	Calm, clear, foggy in the night					
25	42.0	West, light, calm, foggy		25	41.2	Calm, fog, NE, light, ESE, strong					
26	44.3	Calm, NW, fog		26	39.6	WNW, NW, heavy gale, snow, sleet					
27	41.4	West, NW, moderate, fog		27	39.8	Calm, ESE, light, clear					
28	43.2	ENE, light, gloomy		28	43.0	SW, strong, clear					
29	41.6	ENE, strong, misty		29	52.5	SSW, heavy gale					
30	40.3	ENE, fresh, moderate, clear		30	45.6	NW, Heavy gale, rain	Mackenzie River.				
31	42.7	NE, modt. clear, fresh and foggy		31	42.4	Calm, SE, gloomy					
Mean	47.61			Mean	40.85						

NOTE.—The thermometer used in this register, was compared with those in use at Fort Franklin during ten days after our return, and found to coincide with them.

CAPTAIN FRANKLIN'S NARRATIVE RESUMED.

CHAPTER VI.

Brief Notices of the Second Winter at Bear Lake—Traditions of the Dog-ribs—Leave Fort Franklin—Winter Journey to Fort Chipewyan—Remarks on the progress of improvement in the Fur Countries—Set out in Canoes on the Voyage Homeward— Join Dr. Richardson at Cumberland House—Mr. Drummond's Narrative—Arrival in Canada, at New York, and London.

DURING our absence on the sea-coast, Mr. Dease had employed the Canadians in making such repairs about the buildings as to fit them for another winter's residence, but he had not been able to complete his plans before the arrival of Dr. Richardson's party, through whose assistance they were finished shortly after our return. The inconvenience arising from the unfinished state of the houses was a trifle, when compared to the disappointment we felt at the poverty of our store, which contained neither meat nor dried fish, and the party was living solely on the daily produce of the nets, which, at this time, was barely sufficient for its support. Notwithstanding the repeated promises which the Fort hunters and the Dog-ribs in general had given us, of exerting themselves to collect provision during the summer, we found that they had not supplied more than three deer since our departure. The only reason they assigned to Mr. Dease, on his remonstrating with them, was, that they had been withheld from hunting at any great distance from the Fort, by the fear of meeting the Copper Indians, who, they fancied, would be lying in wait to attack them. This excuse,

1826.
September.
Thursday,
21st.

however, had been so often alleged without a cause, that it was considered mere evasion, and we attributed their negligence to the indolence and apathy which mark the character of this tribe.

I need not dilate upon the anxieties which we felt at the prospect of commencing the winter with such a scanty supply of food. We at once sent off five men, provided with nets and lines, to the fishery in M'Vicar's Bay, which had been so productive in the preceding year, in the hope that, besides gaining their own subsistence, they might store up some fish for us, which could be brought to the Fort when the lake was frozen. Our anxiety was, in some measure, relieved on the 28th of September, by the arrival of Beaulieu and some hunters, from the north side of Bear Lake, with a supply of dried meat. The term of Beaulieu's engagement being now expired, he was desirous of quitting our service; and though he was our best hunter, Mr. Dease advised me to comply with his request, as he had collected a number of useless followers, whom we must have fed during the short days. He accordingly took his departure, accompanied by seventeen persons, which was a very important relief to our daily issue of provision. I furnished them with ammunition from the store to enable them to hunt on their way to Marten Lake, where they intended to fish until the return of spring.

Calculating that the stores, which had been ordered from York Factory, must have arrived at Fort Norman, I despatched Mr. Kendall for them; and he returned on the 8th of October, with as much of them as his canoe would carry. The men were immediately furnished with warm clothing, of which the eastern party were in great need, having left every thing on quitting the seacoast, except one suit each. We were rejoiced at the receipt of a large packet of letters from England, dated in the preceding February. They brought out the gratifying intelligence that my friend,

Lieutenant Back, had been promoted, in December, 1825, to the rank of Commander. I likewise received a large packet of newspapers from his Excellency the Earl of Dalhousie, Governor-in-Chief of Canada, to whom I take this opportunity of returning my best thanks for the warm interest he took in the welfare of the Expedition.

I shall now briefly trace the advance of winter: the nights were frosty, and the weather was unsettled and gloomy, from the time of our arrival to the close of September. Heavy rain fell on the 2d of October, which on the following day was succeeded by hard frost and much snow. The snow which fell on the 8th remained on the ground for the rest of the season. The small lake was frozen on the 12th, from which day we dated the commencement of winter, as we had done in the preceding year. There was a succession of gales, and almost constant snow from that time to the close of the month; and on the 30th the thermometer first descended below zero. The snow then was much deeper than at the close of November in the former year. The last of the migratory birds, which were a few hardy ducks, took their departure on the 18th of October.

Stormy weather kept the Bear Lake open until the 16th of November, nine days later than the year before; and for some weeks we received no assistance from the nets, which again reduced our stock of meat to a small quantity. The same occupations, amusements, and exercise, were followed by the officers and men as in the former residence; and the occurrences were so similar, that particular mention of them is unnecessary. On the 25th of November we despatched some men with dogs and sledges to bring the remainder of the stores from Fort Norman. As it was my intention, as soon as the maps and drawings could be finished, to proceed on the ice to Fort Chipewyan, in order to secure provisions for the out-going of the party, and to

2 P

reach England by the earliest conveyance, I requested of Mr. Brisbois to provide a cariole, sledges, and snow-shoes for my journey, the birch of which they are made being plentiful in the neighbourhood of Fort Norman, and he having a better workman than any at our establishment. On the 28th Mackenzie arrived from M'Vicar's Bay, with an acceptable supply of fine white-fish. We learned from him that our party, as well as the Indians, were living in abundance; and that the latter had shown their wisdom this season, not only in taking up their quarters at that place, instead of remaining about the Fort, as they had done in the former year, but also in building themselves houses like those of our men, and thus having more comforts and better shelter than they had ever before enjoyed. The fishery opposite the Fort was now sufficiently productive for our wants, though the fish, from being out of season, disagreed so much with several of the men as to cause great debility, which was the more distressing to us, as we were unable to supply the invalids with meat on more than two days in the week. Contrary to what had happened last season, we did not receive meat this year from more than six or seven persons of either the Hare Indians or Dog-rib tribes, after the ice set in; this happened, probably, from our being now unprovided with goods to exchange for their furs; though they had been expressly told in the spring, that we should have abundance of ammunition, tobacco, and other supplies, to purchase all the meat they would bring.

By the return of our men from Fort Norman, we learned that one of our Dog-rib hunters had murdered a man of his tribe, in the autumn, near the mouth of the Bear Lake River. The culprit being at the house, we inquired into the truth of the report, which was found correct; and he was in consequence instantly discharged from our service. His victim had been a man of notoriously loose habits, and in this instance had carried off the hunter's wife and child, while he was in pursuit of deer, at a great distance from the Fort. The husband pursued the guilty pair the moment he discovered their flight, and, on overtaking them, instantly shot the seducer; but the woman escaped a similar fate, by having the presence of mind to turn aside the muzzle of the gun when in the act of being discharged. She did not, however, escape punishment: her husband struck her senseless to the ground with the stock of his gun, and would have completed her destruction, but for the cries and intreaties of their only child. This transaction adds another to the melancholy list of about thirty murders which have been perpetrated on the borders of this lake since 1799, when the first trading post was established.

The Dog-rib Indians, being derived from the same stock with the Chipewyans, have many traditions and opinions in common with that people. I requested Mr. Dease to obtain answers from the old men of the tribe to a few queries which I drew up, and the following is the substance of the information he procured, which may be compared with the more extended statements by Hearne and Mackenzie, of the general belief of the Chipewyans.

The *first man*, they said, was, according to the tradition of their fathers, named Chapewee. He found the world well stocked with food, and he created children, to whom he gave two kinds of fruit, the black and the white, but forbade them to eat the black. Having thus issued his commands for the guidance of his family, he took leave of them for a time, and made a long excursion for the purpose of conducting the sun to the world. During this, his first absence, his children were obedient, and ate only the white fruit, but they consumed it all; the consequence was, that when he a second time absented himself to bring the moon, and they longed for fruit, they forgot the orders of their father, and ate of the

2 P 2

black, which was the only kind remaining. He was much displeased on his return, and told them that in future the earth would produce bad fruits, and that they would be tormented by sickness and death—penalties which have attached to his descendants to the present day. Chapewee himself lived so long that his throat was worn out, and he could no longer enjoy life; but he was unable to die, until, at his own request, one of his people drove a beaver-tooth into his head.

The same, or another Chapewee (for there is some uncertainty on this head), lived with his family on a strait between two seas. Having there constructed a weir to catch fish, such a quantity were taken, that the strait was choked up, and the water rose and overflowed the earth. Chapewee embarked with his family in a canoe, taking with them all manner of birds and beasts. The waters covered the earth for many days, but, at length, Chapewee said, we cannot live always thus, we must find land again, and he accordingly sent a beaver to search for it. The beaver was drowned, and his carcase was seen floating on the water; on which Chapewee despatched a musk-rat on the same errand. The second messenger was long absent, and when he did return was near dying with fatigue, but he had a little earth in his paws. The sight of the earth rejoiced Chapewee, but his first care was about the safety of his diligent servant, the rat, which he rubbed gently with his hands, and cherished in his bosom, until it revived. He next took up the earth, and moulding it with his fingers, placed it on the water, where it increased by degrees until it formed an island in the ocean. A wolf was the first animal Chapewee placed on the infant earth, but the weight proving too great, it began to sink on one side, and was in danger of turning over. To prevent this accident the wolf was directed to move round the island, which he did for a whole year, and in that time the earth increased so much in size, that all on board the canoe were able to disembark on it. Chapewee, on landing, stuck up a piece of wood, which became a fir-tree, and grew with amazing rapidity, until its top reached the skies. A squirrel ran up this tree, and was pursued by Chapewee, who endeavoured to knock it down, but could not overtake it. He continued the chase, however, until he reached the stars, where he found a fine plain, and a beaten road. In this road he set a snare made of his sister's hair, and then returned to the earth. The sun appeared as usual in the heavens in the morning, but at noon it was caught by the snare which Chapewee had set for the squirrel, and the sky was instantly darkened. Chapewee's family on this said to him, you must have done something wrong when you were aloft, for we no longer enjoy the light of day; " I have," replied he, " but it was unintentionally." Chapewee then endeavoured to repair the fault he had committed, and sent a number of animals up the tree to release the sun, by cutting the snare, but the intense heat of that luminary reduced them all to ashes. The efforts of the more active animals being thus frustrated, a ground mole, though such a groveling and awkward beast, succeeded by burrowing under the road in the sky, until it reached and cut asunder the snare which bound the sun. It lost its eyes, however, the instant it thrust its head into the light, and its nose and teeth have ever since been brown, as if burnt. Chapewee's island, during these transactions, increased to the present size of the American Continent; and he traced the course of the rivers, and scraped out the lakes by drawing his fingers through the earth. He next allotted to the quadrupeds, birds, and fishes, their different stations, and endowing them with certain capacities, he told them that they were in future to provide for their own safety, because man would destroy them whenever he found their tracks; but to console them, he said, that when they died they should be like a seed of grass, which, when thrown into the water, springs again into life. The animals objected to this arrangement, and said, let us when we die be as a stone which, when thrown into a lake, disappears for ever from the sight of man. Chapewee's family complained of the penalty of death entailed upon them for eating the black fruit, on which he granted that such of them as dreamed certain dreams should be men of medicine, capable of curing diseases and of prolonging life. In order to preserve this virtue, they were not to tell their dreams until a certain period had elapsed. To acquire the power of foretelling events, they were to take an ant alive, and insert it under the skin of the palm of the hand, without letting any one know what they had done.

For a long time Chapewee's descendants were united as one family, but at length some young men being accidentally killed in a game, a quarrel ensued, and a general dispersion of mankind took place. One Indian fixed his residence on the borders of the lake, taking with him a dog big with young. The pups in due time were littered, and the Indian, when he went out to fish, carefully tied them up to prevent their straying. Several times as he approached his tent, he heard a noise of children talking and playing; but on entering it he only perceived the pups tied up as usual. His curiosity being excited by the noises he had heard, he determined to watch, and one day pretending to go out and fish, according to custom, he concealed himself in a convenient place. In a short time he again heard voices, and rushing suddenly into the tent, beheld some beautiful children sporting and laughing, with the dog-skins lying by their side. He threw the skins into the fire, and the children, retaining their proper forms, grew up, and were the ancestors of the Dog-rib nation.

On Mr. Dease questioning some of the elderly men as to their knowledge of a Supreme Being, they replied—" We believe

" that there is a Great Spirit, who created every thing, both us
" and the world for our use. We suppose that he dwells in the
" lands from whence the white people come, that he is kind to
" the inhabitants of those lands, and that there are people there
" who never die: the winds that blow from that quarter (south)
" are always warm. He does not know of the wretched state of
" our island, nor the pitiful condition in which we are."

To the question, whom do your medicine men address when
they conjure? They answered,—" We do not think that they
" speak to the master of life, for if they did, we should fare
" better than we do, and should not die. He does not inhabit
" our lands."

On the evening of the 1st of December a brilliant comet
appeared in the western quarter, which had been indistinctly
seen the two preceding nights. A line drawn through α and η
Ursæ Majoris led to its position; it also formed a trapezium with
α Aquilæ and α Lyræ and α Corona Borealis. This was the last
night of its being visible. The temperature had been unusually
high for several days, about this time $+18$ above zero; and, with
the exception of the night of the 1st, the atmosphere gloomy; and
we amused ourselves with conjecturing, whether this extraordinary warmth, and the density of the clouds, could in any way be
ascribed to the comet.

At Christmas we were favoured by a visit from Mr. Brisbois,
to whom we felt much obliged for the care he had taken of our
sea-stores, beside many personal civilities. The visit of a stranger
is always heartily welcomed in such a desolate region, and to provide for the entertainment of the party during Mr. Brisbois's
stay, Captain Back and Mr. Kendall displayed their ingenuity
in cutting out several pasteboard figures, to represent behind an
illuminated screen the characters of a comic piece, which Captain
Back had written for the occasion. The exhibition was entirely

new to most of the party, and its execution afforded such general amusement, that it was repeated on three nights at the request of the men. The New Year was celebrated by a dance, which closed our festivities; and on Mr. Brisbois quitting us the following day, we resumed our ordinary occupations. Two Hare Indians arrived at the Fort, whom Mr. Kendall recognised as the persons who had brought provisions to Dr. Richardson's party, as soon as they had heard of his having reached the Bear Lake Portage; and we had much pleasure in rewarding their promptitude on that occasion, by a substantial present and a silver medal. They were particularly pleased at the medals, and assured us that they should be proud to show them to the rest of their tribe as tokens of our approbation.

On the evening of the 4th of January, the temperature being —52, 2°, Mr. Kendall froze some mercury in the mould of a pistol bullet, and fired it against a door at the distance of six paces. A small portion of the mercury penetrated to the depth of one-eighth of an inch, but the remainder only just lodged in the wood. Much snow fell in the second week of January; and on the 12th, we ascertained that its average depth was two feet in the sheltered parts of the woods. The weather became mild after the 20th; and on the 22d, the sun's rays were so powerful as to raise a spirit thermometer with a blackened bulb, to + 30, 5°. when the temperature of the air was — 3, 5°. A very brilliant and clearly defined parahelion was visible at the time, and there were only a few light clouds. The wind was east, and as usual, with the wind from that quarter when the sky is clear, the distant land appeared much distorted by refraction.

The documents which had been preparing being now nearly finished, we sent for the cariole, &c. from Fort Norman. When the men came back, they brought the information, that, according to the report of the Indians, the ice was so rough on the

Mackenzie above Fort Norman, that travelling would be extremely difficult. I therefore abandoned the intention of proceeding by that way, and resolved on passing through the woods to Fort Simpson, as soon as guides could be procured. The delay afforded me the opportunity of registering the lowest temperature we had witnessed in this country. At a quarter after eight in the morning of the 7th of February, the thermo-meter descended to —58°; it had been —57. 5°, and 57. 3° thrice in the course of this and the preceding day—between the 5th and 8th, its general state was from —48° to—52°, though it occasionally rose to —43°.

At Fort Enterprise, during a similar degree of cold, the atmosphere had been calm : but here we had a light wind, which sometimes approached to a fresh breeze. The sky was cloudless the whole time. Some of our men, as well as the Indians, were travelling on the lake during this cold without experiencing any greater inconvenience than having their faces frost bitten. The dogs, however, suffered severely, three being completely lamed by the frost, and all of them becoming much thinner.* These cold days were followed by windy though mild weather, which brought the rein-deer nearer to the Establishment ; and our hunters killed seven within a day's march. Their re-appearance in our neighbourhood was very gratifying to the whole party, as we were heartily tired of a fish diet, and I felt an especial pleasure at being able to quit the place without the least apprehension of the party being in want of provision.

The following table contains the amount of provision we obtained at Fort Franklin, from the time of Mr. Dease's arrival

* Notwithstanding the severity of the weather, we had great difficulty in causing these animals to depart from their usual custom of sleeping in the snow, and in in-ducing them to occupy the warm houses which were built for them.

to the close of January 1827; independent of the supplies of pemmican, &c. for the sea voyage, which were procured from the Hudson's Bay Company.

Date.	Small Fish. Bear Lake Herring.	Trout.	Fresh Meat. lbs.	Dried ribs of Rein-Deer.	Pounded Deer's Meat. lbs.	Fat or Tallow. lbs.	Rein-Deer Tongues.	Beaver.	Partridges.	Hares.
From 22d June 1825, to 22d June 1826.	49,969	2,313	21,067	1,447	960	627	1,264	9	120	24
From 9th July 1826, to 31st January, 1827.	29,471	1,162	2,986	923	784	2,302	585	3	266	28

On the 16th of February, Augustus and two Dog-ribs were sent forward to be at the track in the line of my intended route. My departure being fixed for the 20th, the charts, drawings, journals, and provisions were distributed between the cariole and three sledges of which my train consisted; and as the dogs were in too weak a condition for drawing heavy burdens, two Indians were engaged, to accompany us four days, for the purpose of carrying part of the pemmican. I afterwards delivered written instructions to Captain Back, directing him to proceed to York Factory as soon as the ice should break, and from thence, by the Hudson's Bay ship, to England, taking with him the British party, but to send the Canadians to Montreal. Augustus and Ooligbuck were to be forwarded to Churchill, that they might rejoin their relatives.

At ten A.M., I quitted the Fort, accompanied by five of our men and the two Indians, the latter dragging each sixty pounds of pemmican on their sledges. Captain Back, the officers, and men assembled to give us a farewell salute of three hearty cheers, which served to renew my regret at leaving a society whose members had endeared themselves to me by unremitting attention to their duties, and the greatest personal kindness. We crossed the lake expeditiously, favoured by a north-west gale, and then continued our course to the southward until sunset. The mode of bivouacking in the winter, as well as the course of proceeding, having been so fully described in my former Narrative, and by several other travellers in this country, I need not repeat them. We usually set forward at the first appearance of light and marched until sun-set, halting an hour to breakfast. The rate of walking depended on the depth of snow; where the track was good, we made about two miles in the hour.

On the evening of the second day, we were deserted by our Indian companions, who, as we afterwards learned, took advantage of the rest of the party being some distance in advance of them, to turn back to the nearest wood, and there deposit the pemmican on a stage which they constructed by the road side. Supposing that they had only halted in consequence of the gale that was then blowing, we did not send to look after them before the following morning, when every trace of their path was covered with the snow drift; and as I considered we might possibly spend some time in a fruitless search, I thought the wisest course was to put the party and dogs on a shorter allowance than usual, and proceed on our journey. Their conduct affords another instance of the little dependence that ought to be placed on the Indians of this country, when more than ordinary exertion is required.

We travelled fifty miles through a swampy level country, thinly wooded, with a few ridges of hills visible in the distance, east and west of our course. The country was uneven and better

February.

March.

wooded for the succeeding thirty miles. We next crossed a steep range of hills elevated about eight hundred feet above the surrounding land, and then passing over a succession of lower hills and vallies, descended to the Mackenzie, and following that river for thirty miles, came to Fort Simpson on the 8th of March; the whole distance being two hundred and twenty miles, and for the last one hundred and seventy miles, through a well wooded country. We crossed several rivers which flow into the Mackenzie, and some considerable lakes which are laid down in the map. But one solitary family of Indians were seen on the journey, and these were stationed within a day's march of Fort Simpson. They had inclosed large tracts of ground with hedges, in which they set snares for hares, and, being very successful, were living in abundance, and were well clothed, their dress consisting principally of hare skins.

As soon as Mr. Smith, the chief Factor of the District, was informed of our approach, and that we were short of provisions, in consequence of the Indians having made off with the pemmican, he kindly sent a supply of fresh meat for our use; and on our arrival at the Fort, he gave us the most friendly reception. Our Indian guide had never been nearer to Fort Simpson by land, than the Lake of the Elevated Land, and only once by the course of the Mackenzie, many years before the Fort was built; and yet if he had not been led aside by falling upon the track leading to the Indians above-mentioned, he would have come upon the Mackenzie directly opposite Fort Simpson. His course he told me was governed by his recollection of a particular mountain, which he remembered to have noticed from the Mackenzie, and which we now passed within two miles, but on his former visit, he did not approach it nearer than eighteen miles. Its outline must have appeared so different when seen from these distances, that one can hardly imagine a less observant eye than that of an Indian recognizing any of its distinguishing points, especially as it was not a detached mountain, but formed one of a line of hills of considerable extent. Our dogs being completely tired, I remained a week to recruit their strength. During this interval, I had the opportunity of examining all the accounts which the Hudson's Bay Company had to present for supplies to the Expedition from this department, and of making provision for the outward journey of Capt. Back and his party. Arrangements were also made, that the Hudson's Bay Company should take, at a valuation, the spare stores of the Expedition on its quitting Bear Lake. I accompanied Mr. Smith to a part of the River of the Mountains, where a portion of the bank, several acres in extent, had been torn off, and thrown a considerable distance into the channel of the river. The disruption took place in the preceding November, some days after the water had been frozen, and when there was no apparent cause for its separation. When the water is flowing over the banks, and the earth is in consequence loosened, the falling of the bank is not unfrequent in the Mackenzie, though on a much smaller scale than in this instance. I can only account for the separation of the mass after the ground had been frozen, by the supposition, that there was some spring of warm water in its rear, which loosened the soil, and that the pressure of the ice contributed, with the weight of snow at the top, to its overthrow.

At the time of my visit, an Indian woman committed suicide, by hanging herself, in a fit of jealousy, at an encampment a short distance from the Fort. I had thought that suicide was extremely rare among the Northern Indians; but I subsequently learned that it is not so uncommon as I had imagined, and I was informed of two instances that occurred in the year of 1826. The weather was remarkably mild; during my stay icicles were formed on the southern front of the house, and there were many other indications of an early spring.

Thursday,
15th.
On the afternoon of the 15th of March I took leave of Mr. Smith, who kindly furnished me with his best dog for my cariole, one of mine having proved unfit for the journey to Slave Lake; we were also indebted to him for the skin of a mountain goat and a lynx; and to Mr. M'Pherson for the skins of several smaller animals and birds, from the neighbourhood of the Rocky Mountains, which they added to our collection. Having sent back one of my men with the Indian guides to Bear Lake, we had now only two sledges; but as we were unable to carry the whole of our lading, Mr. Smith had the goodness to send a sledge and one of his men to convey a part of the provisions for four days. At the distance of eight miles we met two men with a cariole and sledge, which Mr. M'Vicar had sent for my use from Slave Lake; but being well provided, I did not require the services of this party, though we derived great benefit from their track as we proceeded, and also from some deposits of provision which they had made on the route.

Wednesday,
21st.
Following the course of the Mackenzie we arrived, on the 21st, at the expansion of the river called the Little Lake, and there had the pleasure of meeting two Canadians, on their way to Bear Lake, with a packet of letters from England. We hastened towards the shore and encamped; and though the night was piercingly cold, I spent the greater part of it most agreeably, scanning the contents of the box by the unsteady light of a blazing fire. After breakfast next morning I despatched the packet to its destination, under the charge of M'Leay, who had accompanied me from Bear Lake, and retained one of the Canadians in his stead. We arrived at Fort Resolution, on the Slave Lake, to breakfast, on the 26th, and I once more had the happiness of receiving the friendly attentions of

Mr. M‘Vicar, to whom it will be remembered by the readers of my last Narrative, that the members of that Expedition were so greatly indebted for his tender care of them after their sufferings. Dr. Richardson had quitted this place in the preceding December, for the purpose of joining Mr. Drummond, the Assistant Botanist in the Saskatchawan River, and that he might have the benefit of an earlier spring than in this quarter to collect plants. The prospect here being completely wintry, I made another halt of eight days, being desirous of remaining as long as I could, without incurring the risk of exposure to the thaw on my way to Fort Chipewyan.

I was glad to find that the Chipewyans and Copper Indians were at length employing dogs to drag their sledges. A superstitious belief that their own origin was derived from those animals, had for several years past thrown this laborious and degrading occupation on the poor women, who, by the change, experienced a most happy relief. It was, indeed, highly gratifying to observe that these Indians no longer beat their wives in the cruel manner to which they had been formerly accustomed; and that, in the comparative tenderness with which they now treat the sex, they have made the first and greatest step to all moral and general improvement.

It will be recollected that on receiving, at Bear Lake, a report of the traces of white people having been seen near the sea-coast, I had requested that Mr. M‘Vicar would collect a party of Indians, and send them to the spot to convey a letter from me to Captain Parry. Mr. M‘Vicar now informed me that some Indians had left his Fort for the purpose, under the charge of a Canadian, named Joseph St. Pierre, who volunteered for the occasion, but the Indians continued with him only for a short distance beyond the east end of Slave Lake, when they became weary of their journey, and dropping off one by one, left him alone. St. Pierre, however, having determined to deliver the letter to Captain Parry, if possible, persevered for many days in a fruitless search for the river on the banks of which the marks were reported to have been seen; even after he had sustained the loss of all his clothes (except those on his person), by the grass catching fire when he was asleep; but at length, being short of food, his shoes worn out, and almost without covering for his feet, he was compelled to return to the Fort. He was not at the house at the time of my visit, but I left an order with Mr. M'Vicar, that he might be rewarded for his zeal and exertions, and handsomely remunerated for his loss.

April.
The subsequent journey to the Athabasca Lake occupied eight days; we arrived at Fort Chipewyan in the afternoon of the 12th of April. I found Mr. Stewart, the Chief Factor of the Department, surrounded by a large body of Indians, who quitted the Fort as soon as they had exchanged their furs, in order to seek their living by fishing and hunting wild fowl, instead of passing four or five weeks in indolence about the Establishment, as had been their custom at this season for many preceding years. This beneficial change of conduct, on their part, is owing to the Hudson's Bay Company having ceased to bring spirits into the northern department; and to some other judicious regulations which the Directors have made respecting the trade with the natives. The plans now adopted offer supplies of clothes, and of every necessary, to those Indians who choose to be active in the collection of furs; and it was pleasing to learn, that the natives in this quarter had shown their acquiescence in these measures by increased exertion during the preceding winter. Some other very wholesome regulations have been introduced by the Company; amongst others, the Sabbath is ordered to be properly observed, and Divine Service to be read at every post. They have also directed, where the soil will

allow, a portion of ground to be cultivated for the growth of culinary vegetables at each of their establishments, and I witnessed the good effects of this order, even at this advanced post, where the ground is rocky; the tables of the officers being supplied daily, and those of the men frequently, with potatoes and barley. Such luxuries were very rarely found beyond Cumberland House, on the route that we travelled during my former journey.

Feeling a deep interest in the welfare of this country, in which I have spent a large portion of the last seven years, I have much pleasure in recording these improvements; and in stating my conviction, that the benevolent wishes of the Directors, respecting the inhabitants of their territories, will be followed up with corresponding energy by the resident Governor, the chief factors, and the traders of the Company.

I mentioned in my former Narrative, that the northern Indians had cherished a belief for some years, that a great change was about to take place in the natural order of things, and that among other advantages arising from it, their own condition of life was to be materially bettered. This story, I was now informed by Mr. Stewart, originated with a woman, whose history appears to me deserving of a short notice. While living at the N.W. Company's Post, on the Columbia River, as the wife of one of the Canadian servants, she formed a sudden resolution of becoming a warrior; and throwing aside her female dress, she clothed herself in a suitable manner. Having procured a gun, a bow and arrows, and a horse, she sallied forth to join a party of her countrymen then going to war; and, in her first essay, displayed so much courage as to attract general regard, which was so much heightened by her subsequent feats of bravery, that many young men put themselves under her command. Their example was soon generally followed, and, at

2 R

length, she became the principal leader of the tribe, under the designation of the "Manlike Woman." Being young, and of a delicate frame, her followers attributed her exploits to the possession of supernatural power, and, therefore, received whatever she said with implicit faith. To maintain her influence during peace, the lady thought proper to invent the above-mentioned prediction, which was quickly spread through the whole northern district. At a later period of her life, our heroine undertook to convey a packet of importance from the Company's Post on the Columbia to that in New Caledonia, through a tract of country which had not, at that time, been passed by the traders, and which was known to be infested by several hostile tribes. She chose for her companion another woman, whom she passed off as her wife. They were attacked by a party of Indians, and though the Manlike Woman received a wound in the breast, she accomplished her object, and returned to the Columbia with answers to the letters. When last seen by the traders, she had collected volunteers for another war excursion, in which she received a mortal wound. The faith of the Indians was shaken by her death, and soon afterwards the whole of the story she had invented fell into discredit.

In the Athabasca department, which includes Slave Lake and Peace River, as well as in the more southern districts, the autumn of 1826, and the following winter, were unusually mild. Near the Saskatchawan River, there was so little snow before the middle of January, that the sledges could not be used; but at Bear Lake, and throughout the Mackenzie, the weather was severe during the same periods, and the snow came early; hence it would appear, that even in this climate the meteorological register kept at any one place, affords no sure index from whence we can judge of the season at another. In my journey from Slave Lake to the Athabasca we had a snow-storm for three days, which we found did not extend beyond sixty miles; and on our arrival at Fort Chipewyan, we learned there had not been a single shower during these days. The only coinciding circumstance, at the different stations this year, was the prevalence of north-east winds.

We welcomed the appearance of two of the large-sized swans on the 15th April, as the harbingers of spring; the geese followed no the 20th; the robins came on the 7th May; the house martins appeared on the 12th, and in the course of a week were busily employed repairing their nests; the barn or forked-tail swallows arrived on the 20th; and on the same day, the small-sized swans were seen, which the traders consider the latest of the migratory birds.

The only symptoms of reviving vegetation at this period, were a few anemones in flower, and the bursting of some catkins of willows; but we learned by an arrival of a boat from the Peace River that, even so early as the 14th, the trees were in full foliage at not more than a day's journey from the lake. The barley was sown at Fort Chipewyan on the 15th May, potatoes on the 21st, and the garden seeds on the 22d, which were expected to be ready for use by the close of the following September. As an experiment, whether the barley would yield a better crop by remaining in the ground through the winter, some had been sown in the preceding autumn, but only a few of the plants appeared at the close of this month, and the crop did not promise favourably.

Some canoes having arrived on the 26th of May with the furs from Slave Lake, the last of the Company's brigade of boats was despatched to York Factory. Augustus, who was desirous of seeing Dr. Richardson again before his departure from the country, and two other men of the Expedition, embarked in them. I embarked on the 31st May in the Company's light canoe with Mr. Stewart and Mr. M'Vicar, having previously made the necessary arrangements for the passage of Captain Back and his party. We reached Cumberland House on the 18th June, where I had the happiness of meeting Dr. Richardson after a separation of eleven months. I learned from him that during our absence in the north, Mr. Drummond the Assistant Botanist had been indefatigable in collecting specimens of Natural History, having been sent for that purpose to the Rocky Mountains at the head of the Athabasca River; in the course of which service, he had been exposed to very great privations. To his perseverance and industry, science is indebted for the knowledge of several new and many rare quadrupeds, birds, and plants. That the reader may form some notion of the labour he sustained, and the zeal he displayed in making his very valuable and highly interesting collections, and to point out to the naturalist, the districts from whence they were brought, I subjoin his brief account of his journey in his own words.

"I remained at Cumberland House about six weeks after the "departure of Captain Back and Mr. Kendall, in June, 1825, "when the Company's boats with the brigade of traders for "the Columbia, arriving from York Factory, I accompanied "them up the Saskatchawan River two hundred and sixty miles "to Carlton House. The unsettled state of the Indians in that "neighbourhood rendering excursions over the plains very "unsafe, I determined on proceeding with the brigade as far as "the Rocky Mountains. We left Carlton House on the 1st of "September, and reached Edmonton, which is about four hun- "dred miles distant on the 20th of the same month. Sandy plains "extend without material alteration the whole way, and there "is, consequently, little variety in the vegetation; indeed, I did "not find a single plant that I had not seen within ten miles of "Carlton House, although I had an opportunity of examining

1825.
June.

" the country carefully, having performed the greater part of
" the journey on foot.   After a halt of two days at Edmonton,
" we continued our route one hundred miles farther to Fort
" Assinaboyn on the Red Deer River, one of the branches of
" the Athapescow.   This part of the journey was performed with
" horses through a swampy and thickly wooded country, and
" the path was so bad, that it was necessary to reduce the lug-
" gage as much as possible.   I therefore took with me only one
" bale of paper for drying plants, a few shirts, and a blanket ;
" Mr. M'Millan, one of the Company's chief traders, who had
" charge of the brigade, kindly undertaking to forward the rest
" of my baggage in the ensuing spring. We left Fort Assinaboyn
" to proceed up the Red Deer River to the Mountains, on the
" 2d of October ; but the canoe appointed for this service being
" very much lumbered, it was necessary that some of the party
" should travel by land, and of that number, I volunteered to be
" one.   A heavy fall of snow, on the third day after setting out,
" rendered the march very fatiguing, and the country being
" thickly wooded and very swampy, our horses were rendered
" useless before we had travelled half the distance.

October,
2d.

    " We reached the Mountains on the 14th, and I continued to
" accompany the brigade, for fifty miles of the Portage-road, to
" the Columbia, when we met a hunter whom Mr. M'Millan
" hired to supply me with food during the winter.   The same
" gentleman having furnished me with horses and a man to take
" care of them, I set out with the hunter and his family towards
" the Smoking River, one of the eastern branches of the Peace
" River, on which we intended to winter.   My guide, however,
" loitered so much on the way, that the snow became too deep to
" admit of our proceeding to our destination, and we were under
" the necessity of leaving the Mountains altogether, and taking
" up our winter-quarters about the end of December, on the Bap-

" tiste, a stream which falls into the Red Deer River.   During
" the journey, I collected a few specimens of the birds that pass
" the winter in the country, and which belong principally to the
" genera *tetrao* and *strix*.   I also obtained a few mosses, and on
" Christmas day, I had the pleasure of finding a very minute
" *gymnostomum*, hitherto undescribed.

   " In the winter, I felt the inconvenience of the want of my
" tent, the only shelter I had from the inclemency of the weather
" being a hut built of the branches of trees.   Soon after reaching
" our wintering ground, provisions became very scarce, and the
" hunter and his family went off in quest of animals, taking with
" them the man who had charge of my horses to bring me a
" supply as soon as they could procure it.   I remained alone for
" the rest of the winter, except when my man occasionally
" visited me with meat ; and I found the time hang very heavy,
" as I had no books, and nothing could be done in the way of
" collecting specimens of Natural History.   I took, however, a
" walk every day in the woods to give me some practice in the
" use of snow shoes.   The winter was very severe, and much
" snow fell until the end of March, when it averaged six feet in
" depth ; in consequence of this, I lost one of my horses, and
" the two remaining ones became exceedingly poor.   The
" hunter was still more unfortunate, ten of his young colts
" having died.

   In the beginning of April, 1826, setting out for the Columbia
" Portage road, I reached it after a fatiguing march on the sixth
" day, and, two days afterwards, had the pleasure of meeting
" Mr. M'Millan, who brought me letters from Dr. Richardson,
" informing me of the welfare of the Expedition ; and he also
" placed me in comparatively comfortable circumstances by
" bringing my tent, a little tea and sugar, and some more
" paper.   I remained on the Portage preparing specimens of

" birds until the 6th of May, when the brigade from the
" Columbia arrived.   On that day the *Anemone cuneifolia*,
" and *Ludoviciana* and *Saxifraga oppositifolia*, began to flower
" in favourable situations.   My hunter, who had, in the mean
" time, returned to our late wintering ground, now sent me
" word that he had changed his mind, and would not accompany
" me into the Mountains, as he had engaged to do.   His fickle-
" ness deranged my plans, and I had no alternative but to remain
" with the man who had charge of the horses used on the Co-
" lombia Portage, and botanize in that neighbourhood.

" On the 10th of August, I set out with another hunter, upon
" whom I had prevailed to conduct me to the Smoking River,
" although, being disappointed in a supply of ammunition,
" we were badly provided.   We travelled for several days
" without meeting with any animals, and I shared the little
" dried provision which I had with the hunter's family.   On
" the 15th we killed a Mountain sheep, which was quickly
" devoured, there not being the smallest apprehension at the
" time that famine would overtake us—day after day, how-
" ever, passed away without a single head of game of any
" description being seen, and the children began to complain
" loudly; but the hunter's wife, a young half-breed woman,
" bore the abstinence with indifference, although she had two
" infant twins at the breast.   On the 21st, we found two young
" porcupines, which were shared amongst the party, and two or
" three days afterwards, a few fine trout were caught.   We
" arrived in the Smoking River on the 5th of September, where
" the hunter killed two sheep, and a period was put to our
" abstinence, for before the sheep were eaten, he shot several
" buffaloes.

" We proceeded along the Mountains until the 24th of Sep-
" tember, and had reached the head waters of the Peace River,

244

" when a heavy fall of snow stopped my collecting plants for
" that season. I was, however, very desirous of crossing the
" Mountains to obtain some knowledge of the vegetation on the
" Columbia River, and, accordingly, I commenced drying provi-
" sions to enable me to accompany the Columbia brigade, when
" it arrived from Hudson's Bay. I reached the Portage on the

" 9th of October, and on the 10th the brigade arrived, and I
" received letters from Captain Franklin, instructing me to
" descend in the spring of 1827, time enough to rejoin the
" Expedition on its way to York Factory. It was, therefore,
" necessary that I should speedily commence my return, and
" having gone with the brigade merely to the west-end of the
" Portage, I came back again on the 1st of November. The snow
" covered the ground too deeply to permit me to add much to
" my collections in this hasty trip over the Mountains, but it
" was impossible to avoid remarking the great superiority of
" climate on the western side of that lofty range. From the
" instant the descent towards the Pacific commences, there is a
" visible improvement in the growth of timber, and the variety
" of forest trees greatly increases. The few mosses that I
" gleaned in the excursion were so fine, that I could not but
" deeply regret that I was unable to pass a season or two in
" that interesting region.

" Having packed up all my specimens, I embarked on the
" Red Deer River, with Mr. M'Donald, one of the Company's
" officers, who was returning from a long residence on the
" Columbia with his family, and continued to descend the
" stream until we were set fast by the frost. I then left Mr.
" M'Donald in charge of the baggage, and, proceeding on foot
" to Fort Assinaboyn, for the purpose of procuring horses, I
" reached it on the fifth day. It was several days before the
" horses could be obtained, and they were several more in

" travelling from the Fort to Mr. M'Donald, during which time
" that gentleman and his family were very short of provisions.
" The relief, however, arrived opportunely, and they reached the
" Fort in safety. After resting a few days, I set out for Edmon-
" ton, where I remained for some months.

" The winter express brought me a letter from Dr. Richard-
" son, requesting me to join him at Carlton House in April, and
" I accordingly set out for that place on snow shoes, on the 17th
" of March, taking with me single specimens of all the plants
" gathered on the Mountains, lest any accident should happen
" to the duplicates which were to come by canoe in the spring.
" Two men with a sledge drawn by dogs accompanied me, but
" the Indian inhabitants of the plains being very hostile, we
" made a large circuit to avoid them, and did not reach Carlton
" House before the 5th of April. We suffered much from snow-
" blindness on the march, the dogs failed from want of food, we
" had to carry the baggage on our backs, and had nothing to eat
" for seven days. These sufferings were, however, soon forgotten
" in the kind welcome I received from Dr. Richardson, and Mr.
" Prudens, the Company's Chief Trader at Carlton, and the
" hospitable entertainment and good fare of the latter gentle-
" man's table enabled me speedily to recruit my lost strength.

" My collections on the Mountains amounted to about fifteen
" hundred species of plants, one hundred and fifty birds, fifty
" quadrupeds, and a considerable number of insects."

There being yet two months in which Mr. Drummond might
continue his researches, before Captain Back could arrive at Cumberland House, Dr. Richardson had left him on the Saskatchawan
River.

After remaining part of a day at Cumberland House, we
proceeded on our journey, Dr. Richardson following in one of
the Company's boats. I reached Norway House on the 24th of

2 S

June, and Dr. Richardson on the third day after. Mr. Simpson, the resident Governor of the Company, was absent on urgent business at York Factory ; but, previous to his departure, he had provided a canoe, and some additional men, with every other requisite for my journey. We found here Mr. Douglass, who had been sent to the Columbia River by the Horticultural Society, as a Collector of Natural History, and who had recently crossed the Rocky Mountains, for the purpose of proceeding to England from Hudson's Bay. This gentleman being desirous of occupying himself previous to the arrival of the ship, in making an addition to his collection from the neighbourhood of the Red River Colony, I felt happy in being able to give him a conveyance, in the canoe with Dr. Richardson and myself, through Lake Winipeg, to Fort Alexander, where he met another canoe that was going to the colony.

On quitting Norway House we took leave of our worthy companion, Augustus, who was to wait there until Captain Back should arrive. The tears which he shed at our parting, so unusual in those uncultivated tribes, showed the strength of his feelings, and I have no doubt they proceeded from a sincere affection ; an affection which, I can venture to say, was mutually felt by every individual. With great regret he learned that there was no immediate prospect of our again meeting, and he expressed a very strong desire to be informed, if another Expedition should be sent to any of the northern parts of America, whether by sea or land ; and repeatedly assured me, that he and Ooligbuck would be ready at any time to quit their families and their country, to accompany any of their present officers, where-ever the Expedition might be ordered*.

* I have pleasure in mentioning that, by permission of Government, the pay which was due to Augustus and Ooligbuck, has been delivered to the Directors of the Hudson's Bay Company, who have undertaken to distribute it to them annually, in the way suited to their wants.

We reached Fort Alexander on the 8th of July, and Mr. Douglass having left us, I was enabled to offer a passage, as far as Montreal, to Monsieur Picard, one of the clergymen attached to the Roman Catholic Mission at the Red River Colony. We arrived at Lachine, near Montreal, on the 18th of August, and were hospitably entertained by Mr. James Keith, Chief Factor, and Agent of the Hudson's Bay Company, with whom we remained five days, to settle the accounts of the Expedition. After I had paid my respects to his Excellency, the Earl of Dalhousie, Governor in Chief of Canada, we proceeded to New York by the way of Lake Champlain. In our passage through the United States, we received the same kind attentions we had before experienced; our personal baggage, and the collections of Natural History, were forwarded by the officers of the customs without examination, and every assistance we required was promptly rendered.

Having embarked, in the packet ship, on the 1st of September, we reached Liverpool on the 26th, after an absence of two years, seven months and a half. Captain Back, Lieutenant Kendall, and Mr. Drummond, with the rest of the British party, arrived at Portsmouth on the 10th of October. I then received the distressing intelligence of the death of two excellent men, on their homeward passage from Bear Lake to York Factory; Archibald Stewart, who died from consumption; and Gustavus Aird, who was drowned in consequence of his jumping out of the boat, in his exertions to save her, when she was hurrying down the Pelican Fall, in Slave River. Until this account reached me, I had cherished the hope that our Expedition would have terminated without my having to record a single casualty. The loss of these men was the more deeply felt by me, from their uniform, steady, obedient, and meritorious conduct, which I had repeated opportunities of observing and

1827.
July.

August.

September.

2 S 2

admiring, while they were my companions in the Lion, during the voyage along the coast.

I must be allowed to add, that in this long homeward journey, in which there were no fresh discoveries to be made, nor any of those excitements that relieve the monotony of constant labour, and in which they had to contend with a succession of dangerous rapids, there was the same masterly skill and exemplary conduct evinced by Captain Back and Lieutenant Kendall; and the same patient and ready obedience by the men*, which had marked their whole conduct, while more immediately under my own observation.

On my arrival in London, on the 29th of September, accompanied by Dr. Richardson, I had the honour of laying the charts and drawings before his Royal Highness the Lord High Admiral, and Mr. Secretary Huskisson; and, from the latter, I received directions to publish an account of our proceedings.

In concluding this Narrative, I feel it incumbent on me to offer a few remarks on the subject of a *North-West Passage*, which, though it has not been the immediate object of the enterprises in which I have been engaged, is yet so intimately connected with them, as to have naturally excited in my mind, a strong and permanent interest. It is scarcely necessary to remark, that the opinion I ventured to express in my former work, as to the practicability of the passage†, has been considerably

---

* I am happy to add, that those men who had been in His Majesty's service before the present Expedition, have been rewarded by promotion.

† See page 388.

strengthened by the information obtained during the present Expedition. The Northern Coast of America has now been actually surveyed from the meridian of 109° to 149½° west; and again by the exertions of Captain Beechey, in His Majesty's ship the Blossom, from Icy Cape eastward to about 156° west, leaving not more than fifty leagues of unsurveyed coast, between Point Turnagain and Icy Cape. Further, the delineation of the west side of Melville Peninsula, in the chart of Captain Parry's Second Voyage, conjoined with information which we obtained from the Northern Indians, fairly warrants the conclusion, that the coast preserves an easterly direction from Point Turnagain towards Repulse Bay; and that, in all probability, there are no insurmountable obstacles between this part of the Polar Sea and the extensive openings into the Atlantic, through Prince Regent Inlet and the Strait of the Fury and Hecla.

Whenever it may be considered desirable to complete the delineation of the coast of the American Continent, I conceive that another attempt should be made to connect Point Turnagain with the important discoveries of Captain Parry, by renewing the Expedition which was undertaken by Captain Lyon, and which, but for the boisterous weather that disabled the Griper, must have long since repaid his well known zeal and enterprise with discoveries of very great interest.

In considering the best means of effecting the North-West Passage in a ship, it has hitherto been impossible not to assent to the opinion so judiciously formed, and so convincingly stated, by Captain Parry, that the attempt should be made from the Atlantic rather than by Behring's Straits, because the enterprise is then commenced after a voyage of short duration, subject to comparatively few vicissitudes of climate, and with the equipments thoroughly effective. But important as these advantages are, they may, perhaps, be more than balanced by some circumstances which have been brought to light by our Expedition. The prevalence of north-west winds during the season that the ice is in the most favourable state for navigation, would greatly facilitate the voyage of a ship to the eastward, whilst it would be equally adverse to her progress in the opposite direction. It is also well known, that the coast westward of the Mackenzie is almost unapproachable by ships, and it would, therefore, be very desirable to get over that part of the voyage in the first season. Though we did not observe any such easterly current as was found by Captain Parry in the Fury and Hecla Strait, as well as by Captain Kotzebue, on his voyage through Behring's Straits; yet this may have arisen from our having been confined to the navigation of the flats close to the shore; but if such a current does exist throughout the Polar Sea, it is evident that it would materially assist a ship commencing the undertaking from the Pacific, and keeping in the deep water, which would, no doubt, be found at a moderate distance from the shore.

The closeness and quantity of the ice in the Polar Seas vary much in different years; but, should it be in the same state that we found it, I would not recommend a ship's leaving Icy Cape earlier than the middle of August, for after that period the ice was not only broken up within the sphere of our vision, but a heavy swell rolling from the northward, indicated a sea unsheltered by islands, and not much encumbered by ice. By quitting Icy Cape at the time specified, I should confidently hope to reach a secure wintering place to the eastward of Cape Bathurst, in the direct route to the Dolphin and Union Straits, through which I should proceed *. If either, or both, of the plans which I have suggested be adopted, it would add to the confidence and

* See Dr. Richardson's opinion in favour of this route, p. 264.

safety of those who undertake them, if one or two depôts of provision were established in places of ready access, through the medium of the Hudson's Bay Company.

Arctic discovery has been fostered principally by Great Britain; and it is a subject of just pride that it has been prosecuted by her from motives as disinterested as they are enlightened; not from any prospect of immediate benefit to herself, but from a steady view to the acquirement of useful knowledge, and the extension of the bounds of science. Each succeeding attempt has added a step towards the completion of northern geography; and the contributions to natural history and science have excited a general interest throughout the civilised world. It is, moreover, pleasing to reflect that the loss of life which has occurred in the prosecution of these discoveries does not exceed the average number of deaths in the same population at home under circumstances the most favourable. And it is sincerely to be hoped that Great Britain will not relax her efforts until the question of a north-west passage has been satisfactorily set at rest, or at least until those portions of the northern shores of America, which are yet unknown, be laid down in our maps; and which, with the exception of a small space on the Asiatic continent eastward of Shelatskoi Noss, are the only intervals wanting to complete the outline of Europe, Asia, and America.

END OF THE NARRATIVE.

SUMMARY of the DISTANCES TRAVELLED by the EXPEDITION, from its LANDING in AMERICA, until its EMBARKATION.

	Statute Miles.
Distance travelled in 1825, as given in page 50　．　．　．　．	5,803
Dr. Richardson and Mr. Kendall's excursion on the ice to the eastern parts of Bear Lake, in the Spring of 1826　．　．　．　．	359
Distance travelled by the Western Party in 1826 (given in page 285.)　．　．	2,048
Distance travelled by the Eastern Party in 1826, after its separation from the Western Party　．　．　．　．　．　．	1,455
Return from Fort Franklin to New York　．　．　．　．　．	4,000
Captain Back and Lieutenant Kendall's journey to York Factory, after quitting Captain Franklin's route　．　．　．　．　．　．	520
Distance travelled by the Expedition in going and returning, including the excursions of detached parties　．　．　．　．　．　．	14,185

Number of miles surveyed and laid down in the maps, but not all included under the head of discoveries, because the routes have been traversed by Traders　5,000

NOTES on this PUBLICATION

1) This volume is a part of a series of publication on *Arctic discovery*. The volume is published in four parts.
    a. Part 1: Narrative of the Western Detachment by John Franklin, R.N., F.R.S
    b. Part 2: Narrative of the Eastern Detachment by John Richardson. M.D.
    c. Part 3: Thirty one copperplate engravings included in the original 1828 publication (for an example see the next page.)
    d. Part 4: Details shown on six maps of the original publication.

Other Volumes in this series are forthcoming.